# THE WORLD TOMORROW

Photo credits:
Page 16: © NASA; page 23: © NASA; page 51: © Loic Le Quéré; page 72: © Tony Travouillon; page 105: © NASA, SVS, TOM; page 132: © NASA; page 141: © Gérard Lahache; page 160: © Florence Bastien; page 161: © André Monget; page 162: © NASA.


Project Manager, English-language edition: Magali Veillon
Editor, English-language edition: David Bourgeois
Designer, English-language edition: Neil Egan
Jacket design, English-language edition: Brady McNamara
Production Manager, English-language edition: Jacquie Poirier


Library of Congress Cataloging-in-Publication Data

Monget, Yannick.
  [Demain, la terre. English]
  The world tomorrow : scenarios of global catastrophe / Yannick Monget ; preface by Jean-Marie Pelt.
    p. cm.
  ISBN 10: 0-8109-9318-X (hardcover with jacket)
  ISBN 13: 978-0-8109-9318-1 (hardcover with jacket)  1. Global environmental change.
  2. Global environmental change—Pictorial works. 3. Global warming. 4. Global warming—Pictorial works. I. Title.

  QC981.8.G56M6313 2007
  550—dc22
                          2006035492


Originally published in French under the title *Demain, la Terre... Vous ne pouvez pas prédire l'avenir, en revanche, vous pouvez l'inventer.* by Éditions de La Martinière, Paris, 2006.

Copyright © 2006 Éditions de La Martinière, an imprint of La Martinière Groupe, Paris
English translation copyright © 2007 Harry N. Abrams, Inc., New York

Published in 2007 by Abrams Image, an imprint of Harry N. Abrams, Inc.

Printed and bound in the U.S.A.
10 9 8 7 6 5 4 3

harry n. abrams, inc.
a subsidiary of La Martinière Groupe

115 West 18th Street
New York, NY 10011
www.hnabooks.com

YANNICK MONGET

# THE WORLD TOMORROW

## SCENARIOS OF GLOBAL CATASTROPHE

TRANSLATED FROM THE FRENCH BY NICHOLAS ELLIOT

abrams, new york

# preface

Yannick Monget's well-researched volume thoroughly covers the ecological challenges facing us at the dawn of the new millennium. Unlike other books on global warming, *The World Tomorrow* avoids dogmatic statements about future events that remain largely unresolved. By doing so, it reminds us that it is not too late to influence these events and escape fatality.

Monget carefully describes the various global warming scenarios, including alternative ones such as the Gulf Stream's potential deviation from the coast of Western Europe (this would cause significant temperature drops throughout those areas). He highlights the most likely possibilities and supports his statements with precise, accurate data from experts in the field.

Monget insists that in order to fight the now irreversible greenhouse effect, our priority must be renewable, nonpolluting energy sources. Carefully dispensing with "false solutions," such as nuclear power and GMOS (genetically modified organisms), Monget conveys the prevailing opinions of ecologists the world over.

The loss of biodiversity (due to man's disastrous management of the planet rather than to geological accidents), which is commonly accepted today as a sixth phase of species extinction in the history of life, is described as an unavoidable fact. These tremendous changes require man to modify his habits, attitudes, actions, and behaviors with respect to our natural environment as quickly as possible.

The book concludes with suggestions for the kinds of changes required, joining the sadly limited chorus of those attempting to alert their fellow citizens and thwart the perils on the horizon in order to offer future generations a benevolent, welcoming Earth.

*The World Tomorrow* leaves no doubt that Yannick Monget is an exceptionally talented young author. As convincing as he is convinced, Monget clearly has many more tricks up his sleeve. I hope this book will meet the success it so deserves.

—JEAN-MARIE PELT
*President of the European Institute for Ecology and Professor Emeritus, University of Metz*

# table of contents

# introduction

Where do we come from?

Though science remains unable to describe the exact process that led to the evolution of man, our studies of the universe have allowed us to discover its broad outlines. History teaches us that everything probably started with a gigantic explosion some 13.7 billion years ago. The explosion released elementary particles. These particles began to combine in increasingly complex atoms, which developed into the first stars. As stars exploded, they released the primitive dusts that made up the world surrounding us. This process took place over immeasurably long periods of time. Five billion years ago, our planetary system, which consists of nine planets orbiting a central star, was formed. On one of these planets, environmental conditions allowed for a new type of system: life.

Small basic molecules known as proteins brought life to the planet by developing primitive single-cell beings able to reproduce while maintaining increasingly complex and elaborate genetic information.

This is how the first living organisms came into being. Beginning from these common ancestors, life evolved into a boundlessly varied spectrum of living systems and forms. From fish to mollusks, from insects to reptiles—not forgetting birds—a succession of astonishing species colonized our beautiful blue planet. Three and a half billion years later, life gave form to the mammal. Initially too weak to impose themselves in the midst of saurian domination, mammals had to wait tens of millions of years for the ecological niches to open up and for a chance to bring themselves to the fore.

The first primates appeared barely a few million years ago. After a few million more years of evolution, the primates gave birth to our species: *Homo sapiens*. An amazing story ... and an abbreviated, simplified answer to the questions "Where do we come from?" and "Where are we going?"

Maybe not much further...

Today, looking at the most recent studies of our environment, one could certainly be tempted to answer in a downbeat manner. And those 3.8 billion years of evolution on planet Earth are there to remind us that no species is safe from extinction.

Could man be so presumptuous as to imagine he is immortal? Do we have a single argument supporting the theory of such an astonishing capacity for survival? We have only been on this planet for a few thousand years. Is that enough time to ensure our biological perfection? Remember that the dinosaurs dominated the planet for nearly 165 million years before being supplanted by mammals. Remember that ants were building cities hundreds of millions of years before we appeared. Intelligence? Is that what makes us superior? Ironically, that's exactly what seems to be endangering us today. Intelligence provided us with the opportunity to adapt the environment to our lifestyle. Talk about a costly mistake.

Technological comfort is an illusion. Although technology may give us the impression that we can get by without our natural environment, the reality is entirely different. As long as we haven't learned to live totally self-sufficiently in outer space, or to colonize other planets, we will be dependent on our terrestrial environment. Sooner or later, abusing our environment—as we have done for far too many years—will undoubtedly affect us; by destroying our environment, we are condemning ourselves to certain extinction. According to scientific conclusions researchers around the globe have made public in the last few years, the anthropic origin of global warming can no longer be questioned—mankind is directly responsible for the phenomenon.

Throughout the world, voices are being raised to question the organization of our societies, demanding they become more respectful of the environment that shelters us. Though some consider the task Herculean and the goal utopian, it is necessary. Our future depends on it.

The many effects of such ecological turmoil on our civilizations far outstrip what the general public suspects. Though graphs and diagrams may be sufficient to fuel researchers' concerns, they do not speak to the uninitiated. We must advance simulations to the point where every individual can clearly understand the implications of these findings on his or her future.

The images presented here are far from science-fiction fantasies. These are literal translations of various scenarios elaborated by researchers with, among other organizations, the Intergovernmental Panel on Climate Change (IPCC).

Since it is tempting to believe these drastic changes may not affect us, this volume was conceived to do away with that kind of lazy thinking. Wherever we live, whatever our social standing or income, whatever color we are or religion we belong to, we will all be affected. As you will discover throughout this book, global warming is synonymous not only with climate change, but with radical social, economic, and geopolitical change. Yet, however dark the scenarios depicted in these pages may be, far brighter alternatives remain.

Beware: This book makes no claim to predict the future. It only proposes "potential" futures. It is not important to know exactly which changes will take place (no one can say specifically), but to understand that a climate change will take place and that, no matter what it is, it will be damaging to our society's social, political, and economic balance.

Changing the world is not a utopian idea. It is a real possibility and a real solution. Let us not forget that we are the sole instigators of this environmental crisis.

—YANNICK MONGET

THE immediate risk posed by global warming is that the natural disasters with which we have recently been stricken will recur with increasing frequency over the coming years.

The warming of the atmosphere makes our climate increasingly unstable, resulting in more frequent extreme climactic phenomena. In addition, phenomena such as heat waves, floods, and hurricanes will probably become more powerful and more destructive over time.

This chapter will describe what Earth will likely confront in the near future—these events will probably take place in the next few years. Long-term scenarios will be found in subsequent chapters.

## AN EVOLUTION IN CYCLONIC PHENOMENA

While exceptional cyclonic activity made 2004 a very difficult year for areas surrounding the Gulf of Mexico (Appendix 1), 2005 was downright catastrophic: Twenty-six tropical weather systems, including twelve tropical storms and fourteen hurricanes, rose up in rapid succession over the course of a single season (Appendix 2)—seven of which developed into major hurricanes (Appendix 3).

For the first time ever, with the formation of Hurricane Wilma, the list of names ascribed to hurricanes by the Miami Tropical Prediction Center was exhausted in late October 2005. In November of that year, the Center used the Greek alphabet to name five hurricanes—from *alpha* to *epsilon*. Yet the most worrisome fact is not the sheer number of systems recorded but the various phenomena never previously observed or recorded:

- An unprecedented number of tropical systems (twenty-six) was recorded for a single season of cyclonic activity.
- An unprecedented number of hurricanes (fourteen) was observed over a single season.
- A record low in atmospheric pressure (882 hPa [hectopascal]) was recorded over the Atlantic.

A photograph of Hurricane Ivan entering the Gulf of Mexico, taken by astronaut Mike Fincke on September 13, 2004, from the International Space Station.

On the day this record low pressure was logged, climatologists observed a 61-hPa pressure drop in a mere six hours and a 98-hPa drop in twenty-four hours. During this intensification phase, the diameter of the eye of Hurricane Wilma shrank to less than 5 miles (8 kilometers). By way of comparison, in mid-latitude weather systems such as those in Europe, climatologists consider 24-hPa pressure drops over a period of twenty-four hours exceptional (they are referred to as "bombs"). As Swiss weather forecaster Lionel Peyraud put it, "Lothar, a particularly virulent low-pressure system in the midlatitudes, qualified as a 'bomb.' Upon hitting Switzerland in December 1999, its most violent phase caused pressure drops of 20 hPa in six hours and about 40 hPa in twenty-four hours. This was exceptional in several ways, but did not come close to the figures recorded during Hurricane Wilma."

° Finally, for the first time, tropical weather systems formed outside of the usual sectors.

Tropical Storm Delta, for instance, hit the Canary Islands. And on October 9, 2005, Hurricane Victor formed 150 miles (241 kilometers) northwest of the island of Madeira, the farthest north and east a hurricane had ever been recorded over this basin. Hurricane Victor went on to become the first hurricane to reach the shores of the Iberian Peninsula. It should also be noted that Hurricane Victor developed despite relatively low ocean surface temperatures (75°F/24°C) for the formation of a tropical weather system, as well as unfavorable wind altitude.

> Several records have been broken over the last few years: nineteen hurricanes and tropical storms in 1995, sixteen similar events recorded by the U.S. National Hurricane Center in 2003, and twenty-six in 2005.

Is global warming currently causing an increase in cyclonic phenomena? The following paragraphs prove that reality is incredibly more complex. According to Jean Polcher, a French researcher with the French National Scientific Research Center (CNRS) lab for dynamic meteorology, a greatest danger stems from the intensity of these low-pressure systems, which are likely to become increasingly devastating as global temperatures rise.

A closer look at the formation of these storms will give us a more solid understanding of the implications of global warming on the cyclogenesis phenomena.

Hurricanes only form in tropical regions, once summer temperatures have warmed the ocean waters. Cyclonic mechanisms only begin once water temperatures have exceeded 78°F (26°C) over an area 164 feet (50 meters) wide. Under these conditions, water evaporates, allowing hot, humid air into the atmosphere. In the Northern Hemisphere, the rotation of the earth deflects these air masses to the east; in the Southern Hemisphere, to the west. The phenomenon, known as the Coriolis force, provokes a circular movement; the hurricane has been formed.

The driving force behind this phenomenon is warm ocean water. Needless to say, global warming increases ocean temperatures, along with temperatures in the general environment of these intertropical regions.

One could attempt to refute this explanation by stating that global warming is simultaneously causing the glaciers to melt, which has the opposite effect of cooling the earth's bodies of water (a negative retroactive effect). In fact, measurements demonstrate that compared with the effect of global warming, water from melting glaciers has a very small, even negligible, impact on temperature. Water from glaciers is not truly cold (its temperature is at 32°F/0°C when the ice melts), and its volume remains insignificant compared to the total volume of water in the oceans.

Nonetheless, melting glaciers do have an effect on climate. The fresh water from glaciers can modify ocean currents and have a considerable impact on our temperate regions.

> Over the last fifty years, an average of ten cyclonic events (including six hurricanes) have taken place annually in the Atlantic zone, the Caribbean, and the Gulf of Mexico. Since 1995, a significant increase has been recorded: In a single decade, the average has increased from 9.3 cyclonic events to more than 14.
>
> Since 1945, the amount of energy released by hurricanes has increased by 80 percent. Source: Nature, 2005.

One must, however, temper certain radical positions that blame the warming of the earth's waters solely on recent dramatic natural disasters.

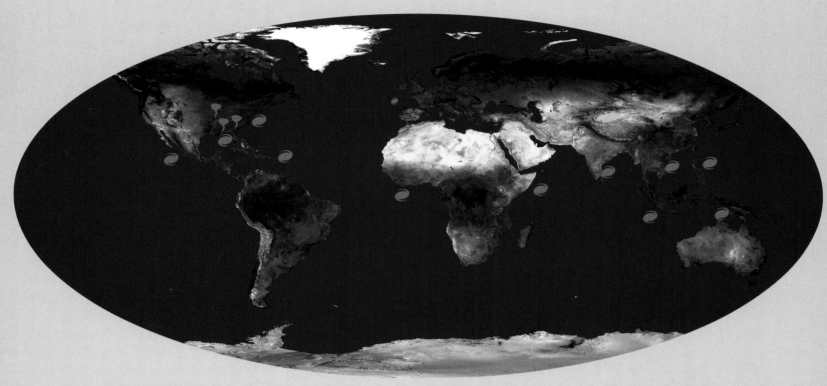

A map of the coasts, cities, and islands threatened by the increase in storms (hurricanes and tornadoes), based on the 1997 IPCC report.

Researchers believe that though a variation in surface water temperatures may have an impact on the formation of future storms, the evolution observed over the last few years is due to a modification in the oceans' thermohaline circulation. Ocean currents are dependent not solely on winds and tides but also on variations in water salinity. Ocean water circulation is therefore composed of horizontal and vertical water currents. This salinity variation is currently evolving in the North Atlantic due to fresh water from the melting of the Arctic ice cap.

It is yet to be determined whether this change is part of a natural cycle, or whether it, too, is an indirect consequence of global warming. (As will be addressed in the following chapter, over the last few years the melting of the polar ice caps has resulted in decreased water salinity, which modifies water density and ocean currents in the North Atlantic.)

Beyond global warming, we are witnessing several interrelated phenomena. The extent to which they are connected remains hard to grasp. The only certainty is that global warming is a reality, and it leads to the warming of the earth's waters.

## VIOLENT TORNADOES STRIKING INLAND AREAS

In North America in 1995, 391 tornadoes were recorded in the United States—at that time, the number set a record for the twentieth century. Since then, the figures have been increasingly worrisome: In 2003, the 1995 record was surpassed when 430 tornadoes were recorded in just a few days (from May 2 to 11). In 2004, this record was pulverized—1,555 tornadoes developed over the first eight months of the year (including 124 over Kansas). The figure dropped in 2005 ("only" 742 tornadoes were recorded), but the number of hurricanes increased, with devastating consequences we won't soon forget.

## MORE FREQUENT HEAT WAVES

Zurich researchers who studied the 2003 heat wave have concluded that global warming could single-handedly account for the phenomenon. Through highly advanced climate simulation systems, the Zurich study demonstrated that global warming leads to greater climate variability, and hence to greater fluctuations from one summer to the next.

Based on an estimated doubling in greenhouse gas emissions, climate models show that by 2041–2100 the average global temperature will have risen by 6°F (3°C); one Swiss model predicts a rise of 9°F (5°C). In these circumstances, climate variability would double from one summer to the next, multiplying the risk of summer drought and causing more frequent intense precipitation.

*While this book was being written [in 2006], NASA confirmed that 2005 was the hottest year since temperatures began being recorded in 1861.*

The year 2005, therefore, beat 1998, the previous record for hottest year on record. However, 1998 was the year of El Niño, which contributed to the rise in overall temperature by 0.4°F (0.2°C). There was no El Niño phenomenon in 2005.

Unlike NASA, the World Meteorological Organization (WMO) does not take the Arctic anomaly recorded in 2005 into account (the Arctic area was particularly hot in 2005) and continues to consider 1998 the hottest year on record. Either way, the two organizations agree on one fact: "Aside from 1996, the last ten years (1995–2005) have been the hottest years ever recorded."

This volume contains several references to positive retroactive phenomena (which aggravate a preexisting phenomenon). A disturbing study undertaken in fourteen sites across Europe during the 2003 heat wave uncovered several phenomena fitting this definition. During the heat wave, Europe, which is generally a carbon sink, became a carbon source due to the considerable slowing of all plant growth (plant growth stabilizes carbon). The effects of this phenomenon were compounded by dramatic forest fires, which traditionally release large amounts of $CO_2$ every summer but were made significantly more violent by the drought.

*Climate simulations predict that, in the future, one out of every two European summers will be as hot as the summer of 2003.*

*European climate studies clearly reveal that since the 1970s an average of fifteen additional days per year have had*

*An awe-inspiring shot of massive forest fires in Northern California [NASA].*

*temperatures exceeding 77°F (25°C). Heat waves are consistently more frequent and are following the general trend of rising temperatures. According to Météo France (the French weather service), heat waves, which are characterized by temperatures of at least 95°F (35°C) by day and at least 71–73°F (22–23°C) by night, will be five times more frequent in France by the end of the century.*

## THE PROBLEMATIC CASE OF EL NIÑO

El Niño stands out among the dreaded short-term effects of these disruptions. Every five years, this meteorological phenomenon takes shape in the Pacific Ocean—a hot current crosses the ocean from east to west, reaching its zenith at year's end along the coasts of the Americas.

The effects of this phenomenon periodically upset the planet's weather balance, with frequently catastrophic results. The exceptionally powerful effects of the 1997 El Niño phenomenon have gone down in history. Torrential rains battered South America, pouring more than fifteen times the average annual precipitation onto the coasts of Ecuador and Peru. Downpours, floods, landslides, and, eventually, cholera and malaria epidemics devastated the region. Yet across the ocean, El Niño brought drought to Indonesia and Papua-New Guinea, where fires destroyed more than 5 million acres (2.02 million hectares) of forest.

Since 1976, occurrences of El Niño have been atypical. For the last twenty years, El Niño seems to have increased both in power and in frequency, which naturally leads us to speculate about the link between this phenomenon and global warming.

Climate models show that rising temperatures could lead to a more intense El Niño. This argument is supported by a variety of studies that reveal indisputable anomalies by comparing the most recent phenomena with weather conditions in previous centuries. Yet here again, researchers do not yet have enough distance, and models provide widely varied results. As with cyclogenesis phenomena, we will only be able to confirm these fears or put them in perspective by studying the next occurrences of El Niño.

# OVERALL, WHAT CAN WE EXPECT OVER THE NEXT FEW YEARS?

Though the evolution of El Niño and tropical storms remains subject to tremendous uncertainty, a certain number of changes—summarized below—are likely to occur in the near future.

### AFRICA
- Increasingly frequent droughts throughout all of the Mediterranean regions and in the south of the continent.
- Potential acceleration of desertification in the north and south.
- Droughts and floods in other regions.

### OCEANIA
- Probability of more intense droughts and tropical storms.

### ASIA
- Increasingly frequent extreme weather events in temperate and tropical areas (hurricanes, floods, droughts, forest fires).
- Hurricanes and rising sea levels along Asia's tropical coasts, possibly driving millions of people inland.

### SOUTH AMERICA
- Increasingly frequent floods, storms, and hurricanes.
- Rising sea levels, threatening many coastal areas.

### NORTH AMERICA
- East Coast: increasingly powerful hurricanes.
- Central: increasingly frequent droughts.

### EUROPE
- Northern: more significant rain in the winter, leading to increasingly frequent flooding.
- Southern: more significant droughts.

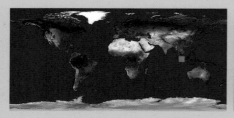

Southeastern Asia could be increasingly affected by violent weather. At right, downtown Singapore at the mercy of violent storms.

"No, everything is not indefinitely possible. Space is not extendable, time itself is limited. If we're not careful, one day mankind will no longer hold the cards."

—**Nicolas Hulot**, *Le Syndrome du Titanic* (The Titanic Syndrome), 2004

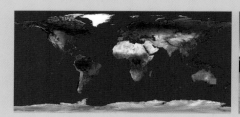

Sydney, Australia. Fires reach the coast and force people to evacuate by sea. Canberra recently faced a similar situation: Several neighborhoods in Australia's capital were totally leveled by fires in January 2003.

*"The challenge we face is nothing less than to ensure the survival of humanity."*

—Mikhail Gorbachev

"Do not wait for extraordinary circumstances to do good; try ordinary situations."

—Johann Paul Friedrich Richter

A storm over Chicago, Illinois. Due to global warming, tornadoes could become increasingly violent over the Midwest, the area of the world most frequently exposed to the formation of these meteorological systems.

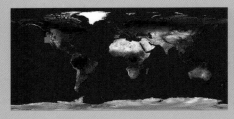 

California devastated by fires. Like certain areas of Australia or Mediterranean Europe, California will face increasingly hot, dry summers. In this picture, fires ravage part of downtown San Diego after California authorities were unable to control gigantic forest fires.

*"Our house is burning and we're looking the other way. We refuse to admit that mutilated, overexploited nature can no longer manage to regenerate. Humanity is suffering. It is suffering from poor development, in the north and in the south, and we are indifferent. Earth and humanity are in danger and we are all responsible."*

—**Jacques Chirac**, Johannesburg Summit, September 2, 2002

"*We are experimenting with climate on a global scale....
Unlike the person carrying out an experiment, we
cannot stop the experiment in case it goes bad...we
are in the test tube. Not only us, but our children and
our grandchildren.*"

—**Hubert Reeves**, *Mal de terre* [Earth Sickness], 2003

Torrential downpours unleash terrible floods in Central Europe, including in
Prague. Floods, which already devastated these areas in 2002, will probably
become more severe in the near future.

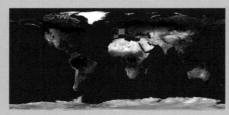

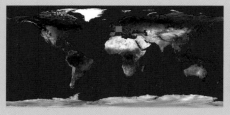

A new heat wave strikes Europe. Here, the deserted lawn of the Reichstag in Berlin has been replaced by dusty, cracked earth. Heat waves similar to those of 2003 are likely to affect these countries with increasing frequency.

"*Protecting the environment is expensive, but the cost of failure is even greater.*"

—**Kofi Annan**, Opening Speech, Earth Summit, September 2002

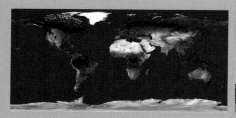

A violent monsoon batters Southeast Asia, hitting many countries with torrential downpours, causing floods and landslides such as those seen here in Bombay, India, at the Gateway of India.

"True wisdom consists not in seeing what is right before our eyes, but in foreseeing what is to come."

—Terence

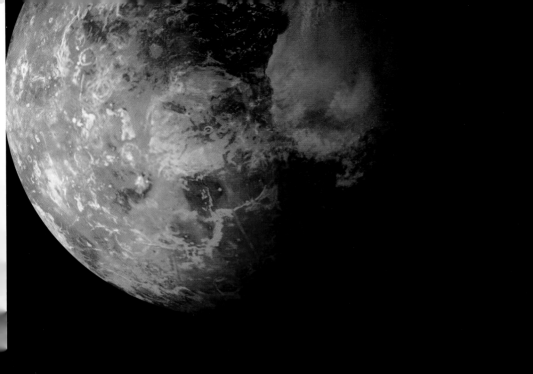

La Défense Development, Paris

Leaning Tower of Pisa, Italy

Big Ben, London

Eiffel Tower, Paris

Les Invalides, Paris

Mount Kilimanjaro, Tanzania

Space Needle, Seattle

Ruins of the Forum, Rome

Palácio del Congresso, Buenos Aires

# RISING WATER

$\mathrm{T}$HE second consequence of global warming is that water levels around the world are rising. The danger is undeniable, as sea levels have already started to rise, gnawing away at various coasts in Europe, Asia, Africa, and the Americas. In Alaska, for instance, several villages have been moved, while in the Indian and Pacific oceans, entire islands have been evacuated, and many archipelagos, such as the Maldives, now only rise a few inches from the ocean surface. (This explains why entire islands were wholly submerged during the devastating tsunami that hit Southeast Asia in December 2004.)

## SEA LEVELS ARE RISING

The increase in the average elevation of ocean levels is not due to the melting of polar ice, but to global warming, which as it warms the oceans, expands their volume of water (warm water occupies more space than cold water). (The melting of polar ice has no effect on the rise in ocean water; only the melting of land glaciers—in the Himalayas, the Alps, etc.—have an impact on sea levels. However, as will be addressed in the following chapter, melting glaciers have other effects, principally on the climate of temperate regions.) The rising temperatures currently dilating the oceans are constant throughout the earth's significant bodies of water.

## ALREADY FIVE TO EIGHT INCHES

According to the IPCC's third evaluation report, sea levels rose 5 to 8 inches (13 to 20 centimeters) during the twentieth century. The report raises the alarming possibility that this phenomenon (which is rapidly accelerating) could affect the water cycle and the availability of drinking water, as well as land, fresh water, and coastal ecosystems.

*Forty-four percent of all floods between 1987 and 1997 took place in Asia. They caused 228,000 deaths and an estimated $143 billion in damage.*

A simulated satellite image of Europe after the glaciers have melted (with a total rise in sea level of 260 feet/79 meters). Brittany has become an island; the Paris Basin is completely underwater; and Marseille, Bordeaux, and London have been wiped off the map. (Note that the peaks of the Alps and the Pyrenees are devoid of snow.)

## ENTIRE POPULATIONS AT RISK

A billion people, or about one sixth of the earth's current population, inhabit areas threatened by centenary floods. According to experts with the United Nations University, if nothing is done to monitor the climate changes underway, with the rise in sea levels and expanding deforestation in areas prone to floods (whose rich soil tends to attract settlers), the situation by 2050 will be considerably worse. By then, 2 billion people (one fifth the global population forty years from now) will be living in areas threatened by floods.

> *Large populations are threatened with exodus due to rising sea levels in several areas: 15 to 20 million people in Bangladesh, 10 million in the Mekong Delta, 4 million in Nigeria—not to mention those threatened by rising water levels in Nepalese lakes due to ice melting in the Himalayas.*

> *More than 2 billion people live at less than 10 feet above sea level.*

## SEVERAL COUNTRIES ARE FATED TO DISAPPEAR OVER THE NEXT FEW YEARS!

Due to rising sea levels, the evacuation of the Tuvalu archipelago, which lies halfway between Hawaii and Australia, has already been planned. This one-time British colony, which has been independent since 1978 but is still a member of the Commonwealth, covers a surface of 10 square miles of land (27 square kilometers) spread over nine islands, and 270,270 square miles (700,000 square kilometers) of the Pacific Ocean. It is the fourth-smallest country in the world. Maximum elevation: 15 feet (4.5 meters) above sea level.

Lester Brown, president of the Earth Policy Institute, has explained that the decision to evacuate Tuvalu was made due to the growing threat of floods. Floodwaters have been reaching 10.5 feet (3.2 meters) in an archipelago whose highest point is only 15 feet (4.5 meters) above sea level. "Tuvalu's leaders have conceded the battle against rising sea levels," said Brown, and ordered the evacuation of the 11,000 citizens inhabiting the nine Tuvalu atolls. Although Australia has refused shelter to the islands' environmental refugees, New Zealand has been accepting them since 2002. But what will happen, Brown wonders, to the 311,000 inhabitants of the Maldives, fated to face a similar situation in the coming decades? The famous Tonga and Clipperton islands in the Pacific Ocean (average elevation of only 6.5 feet / 2 meters above sea level) have also been affected. Having factored in the erosion of the limestone crown, scientists agree the atoll will be entirely underwater in two hundred years.

According to French Senate Report #355, dated May 25, 2000, the disappearance of many island states in the Caribbean, the Pacific, and the Indian oceans will have serious consequences. Without recourse in the face of this imminent threat, a group of about forty island states such as Tuvalu have finally joined forces in the last few years in order to raise awareness of their plight and attempt to sensitize the United Nations to the danger of their total submersion. Recently, the prime minister of Tuvalu, Saufatu Sopo'aga, confirmed that his government may pursue legal action against Australia and the United States for their emission of greenhouse gases, considered responsible for global warming.

## INEVITABLE DISASTERS

Sadly, no matter how much we reduce our greenhouse gas emissions, these disasters can no longer be avoided. The process has significant inertia, which means it will probably take several thousands of years for the situation to stabilize. Over the coming decades, and possibly even centuries, the earth's sea levels will continue to rise, a process that will be compounded by the planet's increased temperature.

## A VARIATION OF "JUST" A FEW DEGREES

In the twenty-first century, scientific estimates of the rise in the world's temperature range from 2.5 to 10.5°F (1.4 to 5.8°C). By comparison, a drop of 9°F (5°C) during the last glaciation led to a roughly 390-foot (119-meter) drop in sea levels. (One could have walked from France to England, as both Europe and Canada were covered by giant glaciers.)

A small variation in temperature on a global scale can have devastating effects on our climate. The resulting rise in sea levels could lead to a radical, planetary change in the earth's geography by submerging vast coastal regions on every continent. As an example, nearly 20 percent of the total surface area of Bangladesh is at risk.

## A THEORETICAL 275-FOOT ELEVATION IN SEA LEVELS

This chapter features images of what the planet would look like if the glaciers were to melt completely. What should we expect if such a catastrophe takes place?

In order to answer this question, we must first remember that the estimated total volume of continental ice is estimated at 7 million cubic miles (29 million cubic kilometers), with over 95 percent in Antarctica. The oceans' surface, which covers about 70 percent of the planet, is equivalent to about 138 million square miles (357 million square kilometers). If the glaciers were to melt entirely, sea levels would rise by 0.05 of a mile, or 275 feet (84 meters).

In reality, this figure will be slightly lower because a small portion of Antarctic ice does not adhere to the continental shelf, which is below sea level in this area. (Remember, thawing ice chunks in the oceans will not modify sea levels—fresh water running into the ocean will occupy the volume of sea water initially occupied by ice.) The theoretical sea level would therefore reach about 249 to 255 feet (76 to 78 meters). Though it may seem completely incredible to us, it is theoretically possible that Paris or London could be entirely swallowed up by the Atlantic in the foreseeable future.

In France, many strips of land have already disappeared, both on the Atlantic and along the Mediterranean, where the Camargue is currently paying the highest price for rising sea levels. For several decades now, the sea has been gaining 0.07 of an inch (0.18 of a centimeter) in the mouth of the Rhône River. Aside from the lagoons of the Camargue, in the coming years the Gironde estuary should also be affected by rising sea levels, for the advance of salt water in areas near estuaries will threaten potable water supplies and agriculture.

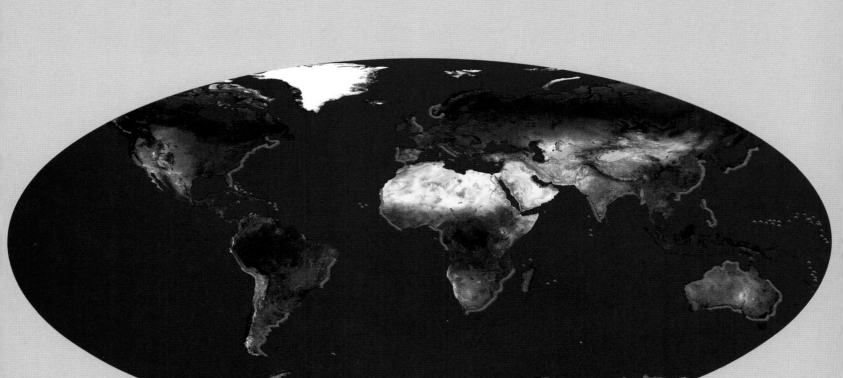

Map of coastlines and islands likely to be submerged by a rise in sea levels (based on the 1997 IPCC Report).

According to the IPCC, the rise in sea levels is estimated to be from 5.9 to 37.4 inches [15 to 95 centimeters] over the next decades.

According to researchers, a 3-foot [1-meter] rise in the sea level will cause coastlines to recede about 300 feet [91 meters].

## MELTING GLACIERS

Though it only partially contributes to the rise of sea levels around the globe, the melting of the continental glaciers does pose an imminent threat for vast populations. The entire Indian subcontinent is endangered by the increasingly speedy melting of the 15,000 glaciers in the highest mountain chain in the world, the Himalayas. The glaciers are melting so rapidly that experts estimate they may be entirely gone from the Himalayas within thirty-five years!

The immediate consequences (in the coming years) are terrifying. Aside from the risk of high-altitude lakes causing landslides in many inhabited areas, the additional water reaching the valleys could make the Ganges and Indus rivers overflow, flooding an entire section of the Indian continent. Ironically, in the long term, the disappearance of the glaciers will eventually dry up these two rivers that currently provide water to more than 500 million people.

IN ALASKA
According to the magazine Science, researchers with the Geophysical Institute at the University of Alaska [Fairbanks] have established the following facts: 85 percent of glaciers are melting at an alarming rate [much faster than we imagined]. Two glaciers are melting at a particularly rapid pace: the

Columbia Glacier, which is shrinking by 26 feet (8 meters) a year, and the Bering Glacier, which is shrinking by 9 feet (3 meters) a year.

## ON MOUNT KILIMANJARO

Several dozen feet of snow each year disappear from Kilimanjaro. In just a few years, the snows of Kilimanjaro will be a thing of the past.

## IN THE HIMALAYAS

Sitting to the east of the mountain chain, the Dokriani Barnak Glacier, which was 3 miles (5 kilometers) long in 1990, is nearly 0.5 miles (0.8 kilometers) shorter than it was ten years ago.

## IN THE ALPS

According to a recent study by Météo France's Center for the Study of Snow (CEN) in Grenoble, the temperature in the French Alps since 1958 has increased by 2°F to 6°F (1°C to 3°C)— significantly more than in the rest of the country.

## IN GREENLAND

Between 2000 and 2005, the Helheim Glacier (one of the largest in the Arctic) was observed to retreat by 4.4 miles (7 kilometers), fueling fears that the Greenland ice cap may disappear twice as quickly as expected (these ice caps contain enough water to make global sea levels rise by 13 to 19 feet/ 4 to 6 meters).

## THE MELTING OF THE POLAR ICE CAPS—ENTIRE ECOSYSTEMS AT RISK

Other areas of the world affected by melting ice include the Arctic and Antarctic polar ice caps. Aside from the changes brought on by large volumes of fresh water spilling into the oceans, the melting of the ice caps will lead to the disappearance of an entire land and marine ecosystem.

Among the endangered species, the polar bear may be extinct by the end of the century. (Its survival would be particularly difficult if, as certain climate models predict, ice floes were to melt during the summer months.) Sadly, the polar bear's extinction is only the tip of the iceberg (no pun intended!). Many other species would likely become extinct once all the ice had melted. The marine ecosystem itself would be severely affected. For instance, researchers have pointed out that krill banks (the basis of countless food chains) can only be found "from 0.6 to 8 miles (1 to 13 kilometers) from ice coasts."* This gives us a better idea of the disastrous impact of these zones' disappearance on the area's environment, but also on many other species such as baleen whales, which feed primarily on small crustaceans.

*According to Dr. Andrew Brierley (Project Manager for the British Antarctic Survey)

The polar bear, threatened with extinction just like thousands of other Arctic species, may be doomed to survive in captivity, scattered in a few zoos around the planet. Aside from the destruction of its ecosystem, the polar bear must also face increasingly high incidences of cancer due to water pollution in the world's oceans.

The La Défense development in Paris. One by one, the great nations have collapsed. Since France doesn't have the means to build massive dams upriver, the entire Paris Basin is submerged by water from the Atlantic Ocean.

*"Those of us who revel in nature's diversity, those of us for whom each animal is a master capable of expanding our knowledge, tend to consider the arrival of the Homo sapiens as the greatest disaster since the extinction of the Cretaceous."*

—**Stephen Jay Gould**, *The Panda's Thumb: More Reflections in Natural History*, 1980

*"Blessed are the meek, for they will inherit the earth."*

—Matthew, 5:5

The city of Pisa, which sits close to the coast, is overtaken by the warm waters of the Mediterranean.

"*The cities of the Gentiles fell. And great Babylon came in remembrance before God, to give her the cup of the wine of the indignation of his wrath.*"

—The Apocalypse of Saint John, 16:9

In this photograph of London, Big Ben emerges from the water eroded by the waves of the Atlantic, which has permanently melded with the Channel.

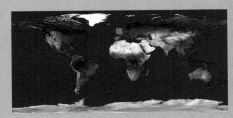

Another photo of Paris under water. Here, the Champ de Mars and the Eiffel Tower [seen from the Trocadéro]. As water levels rise throughout the Paris Basin, salt gradually withers all plant life. In a few weeks, all the trees in the flood zones will have permanently lost their leaves.

"We are sailing through troubled waters that are presumably becoming increasingly hot."

—**Hubert Reeves**, *Mal de Terre* [Earth Sickness], 2003

Another picture of Paris. Here, the Invalides.

"When there is a risk of severe or irreversible perturbations, the absence of absolute scientific certainty must not serve as a pretext to defer the adoption of measures."

—Precautionary principle defined by the U.N. in 1994

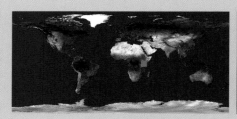

Mount Kilimanjaro in Tanzania, still known as the Uhuru Peak, rises 19,345 feet (5.9 kilometers) above the Kenyan savannah but has lost its legendary eternal snows. The melting of land glaciers will be largely responsible for the rise of global sea levels. In the case of Kilimanjaro, this should be a reality in the very near future, possibly this very year in the worst-case scenario, and certainly by the end of the decade.

"Stop dancing on the edge of the volcano."

—Proverb

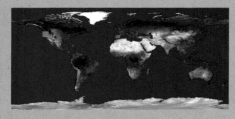 

Seattle lies in ruins, battered by waves from the Pacific Ocean.

"We are nature. But nature is also everything that resists man's will and holds him hostage."

—Jean-Louis Étienne,
*Le Pôle intérieur* [The Interior Pole], 1999

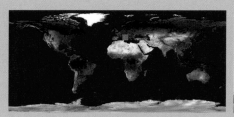

The ruins of Rome. Dams built to hold back the waters of the Mediterranean from the city have given out, worn down by the passage of time.

*"How sad to think nature speaks and mankind doesn't listen."*

—**Victor Hugo**, *Océan, Tas de pierres* [Ocean, Pile of Stones], 1839

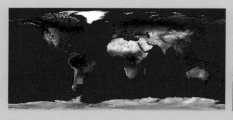

The ruins of Buenos Aires. Here, the Palácio del Congresso, the Argentine parliament, barely rises out of the waters of the Atlantic Ocean.

"Nature, to be commanded, must be obeyed."

—Francis Bacon, *Novum Organum*, 1620

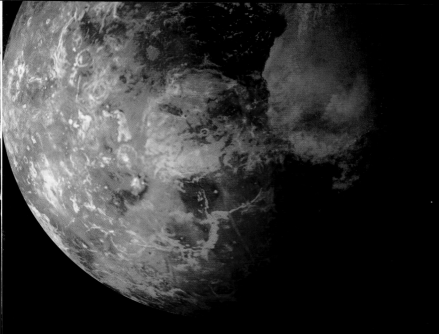

Island of Manhattan, New York

Saint Basil's Cathedral, Moscow

Tower Bridge, London

Atomium, Brussels

Munich Glyptothek, Germany

Château de Chambord, France

Mont-Saint-Michel, France

Metz train station, France

# THE NEW ICE AGE

THE idea that a change in one of the parameters governing the natural balance of the earth could play havoc with our environment is a particularly ancient one. The philosophical intuitions of Greek thinkers such as Plotinus or Eastern thinkers such as Buddha, both of whom spoke of "universal sympathy" and "interdependence of phenomena," are remarkably similar. Today, at the dawn of the third millennium, science is making its own contribution and demonstrating that these intuitions will probably become realities in the near future.

The scenario considered over the next few pages is caused by a phenomenon already underway. It has been observed in the glacial ice of the Arctic Ocean since the 1970s. For several decades, Arctic ice has been melting, pouring large volumes of fresh water into the ocean. We referred to this phenomenon in previous chapters, noting that the melting of the ice floes would not cause sea levels to rise (contrary to the melting of land glaciers). However, as described below, the drop in salinity in the ocean* is likely to have catastrophic secondary effects on the changing temperate climates of Europe and North America (with the rise in local temperatures and increasingly frequent rainfall serving as aggravating factors).

*Water contained in polar ice is fresh water. Unlike ocean water, it does not contain salt. The melting of the polar ice cap therefore leads to a drop in the Arctic Ocean's salinity.

Submarines passing under the North Pole have recorded a 40 percent drop in the thickness of the ice floes.

By 2100, the Arctic would be an open ocean. The local temperature in the Arctic has already increased by 4.5°F [2.5°C] in fifty years.

## THE EFFECTS OF THE DROP IN SALINITY ON THE GLACIAL OCEAN

In order to better understand the phenomenon, we should first recall that the earth's poles receive less solar energy than areas along the equator, leading to a temperature difference that provokes a constant flow of air and water between the poles and the equator.

Polar ice melting: tipping point of global warming.

The Earth's rotation interferes with these currents, resulting in cyclones (see chapter 1) and anticyclones or ocean currents such as the Gulf Stream. Ocean currents are responsible for the temperate climate currently enjoyed in Europe. Paris is, after all, farther north than Montreal, on the same latitude as Quebec City.

The Gulf Stream functions as a powerful source of heat (equivalent to over a million power stations), moving warm tropical water along the shores of North America to the coasts of Spain, France, Britain, and Norway.

The drop in the oceans' salinity constitutes a threat to the climatic balance of all temperate zones. By slowing the sinking of polar water to the bottom of the sea, it modifies currents and could potentially put a halt to the Gulf Stream.

The phenomenon is growing with every passing month. In fact, the disaster could be upon us more quickly than anyone expects: A slowing of the Gulf Stream has already been recorded. The phenomenon is underway. The only thing scientists remain in the dark about is when natural Arctic equilibriums will be broken, permanently ending the Gulf Stream.

## A DRAMATIC, VICIOUS CYCLE

The melting of permanent ice has already drawn us into a vicious cycle, the effects of which are snowballing. The ice-free ocean absorbs more solar energy than ice, which is highly reflective, leading to an acute increase in water temperature, which serves to accelerate the melting of the ice and perpetuates the cycle.

A similar phenomenon can be observed in the frozen expanses of the Siberian and Canadian permafrost. Since the extent of the snow cover is diminishing, the ground no longer reflects sunlight, but absorbs it. As it warms up, the permafrost releases methane (a greenhouse gas far more powerful than $CO_2$), which also contributes to the warming of the atmosphere.

This increase in polar temperature is all the more rapid in that the high latitudes are very sensitive to the increase in $CO_2$. Indeed, air in the high latitudes is colder and contains less vaporized water, which generally serves as a buffer for temperature change. These areas are therefore far more sensitive to climate variations.

On a larger scale, once global warming passes a certain threshold, it may well become self-sustaining. Several so-called positive retroactive phenomena are involved.

Though it took human technology to release large quantities of the greenhouse gases contained in fossil fuel reserves (e.g., oil, coal), it is feared that global warming will soon evolve without human assistance.

As we have seen, greenhouse gases exist in various forms and are stored in a variety of environments. The permafrost could lose between 15 and 30 percent of its surface in the next fifty years, releasing a significant volume of methane. Ocean plateaus could release major amounts of clathrates (gas hydrates implicated in climate change). The spongy terrain of peat bogs threatens to dry out, exposing large quantities of organic matter to the atmosphere, which would lead them to degrade and release large amounts of $CO_2$. As oceans—Earth's principal sources of carbon—warm up, they will dissolve less atmospheric $CO_2$. Heat waves in Europe have also taught us that vegetation may cease to store carbon during droughts.

If this should happen, mankind, which set global warming in motion, would have no way of stopping or reducing it. And the dramatic, vicious cycle would be underway.

American and Canadian researchers have announced that the most important ice shelf in the Arctic, which is over three thousand years old, ruptured between 2000 and 2002. They believe this occurred due to global warming.

Earth has lost 10 percent of its snow-covered surface in forty years.

The permafrost is melting so quickly that buildings in various Alaskan cities have collapsed.

## THE RISKS OF A SUDDEN CLIMATE CHANGE

*"The Arctic is changing rapidly. We should be concerned that this is happening now and that we will have to adapt to this change."*

—**Mark Serreze**, researcher at the University of Colorado

The end of the Gulf Stream could cause brutal climate change. The latest readings from Nordic regions indicate that the change could happen very soon, probably before the end of the century. It is therefore entirely plausible to imagine that within fifty years this sea conveyor belt could come to a total halt. If this were to happen, areas around Europe would become considerably colder.

Though it seems unlikely, this scenario could be even more pronounced and could take place even more quickly. In fact, the change could happen within twenty years and be far more violent. In this case, the south of England would be covered in a sea of ice, and violent blizzards and freezing rain would reach far into the European mainland. The infrastructure of European nations would not withstand this extreme weather. Electric cables would freeze, depriving vast areas of electricity. If this scenario were to come true, there would not be enough time to issue any warnings.

This phenomenon could serve as a catalyst for other changes and have a global impact: While Europe would be subjected to a kind of ice age, the Tropics would move lower. Experts believe that rains in Central America and in humid regions such as the Amazon would diminish by about 40 percent, modifying ecosystems and transforming the equatorial jungle into vast prairies. In Asia, the monsoon would be eradicated. Droughts could provoke a reduction in methane, which would cause another temperature drop. Throughout the world, famines would occur more frequently, leading to various disturbances (the exodus of tens of millions of people, etc.) that would probably be fatal to our economies.

## SCIENTISTS CAN PROVE IT—IT HAS HAPPENED BEFORE

Unfortunately, recent discoveries have proved that a similar event took place only a few millennia ago.

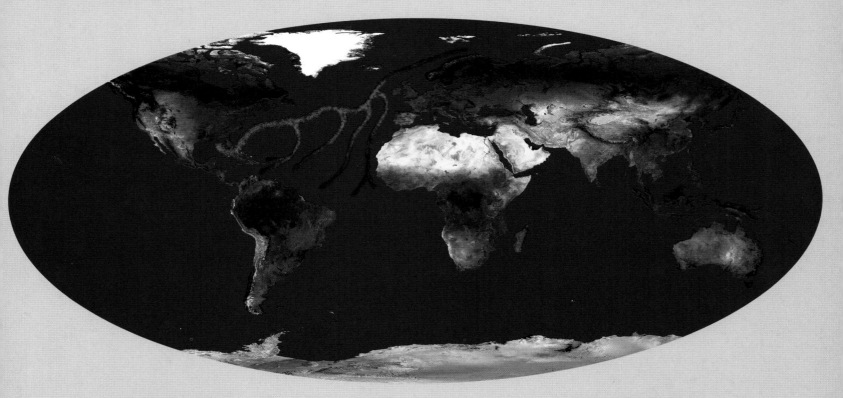

The Gulf Stream forms in the warm waters of the Gulf of Mexico, passes between Cuba and Florida, and travels northward along the coast of Florida. This is where the current is the strongest [50 miles/80 kilometers out to sea, 2,100 feet/640 meters deep, moving at a speed of 60 to 90 miles/97 to 145 kilometers a day, a water temperature of 86°F to 95°F/30°C to 35°C, and a rate of flow of 3 billion cubic feet/85 million cubic meters of water a second]. Farther north, the Gulf Stream is repelled by the cold waters of Labrador and loses much of its speed [4.9 miles/7.9 kilometers a day] as it heads toward Europe and divides into two currents: one moving toward Iceland, and one toward the Azores. Along the coast of Norway, the current's water temperature has dropped to 4°F [2°C], and its density is much higher due to high salinity levels caused by evaporation. At this stage, the Gulf Stream waters sink toward the bottom of the ocean.

Paleoclimatologists have long been aware of a sudden climate change some eighty-two hundred years ago. Measurements showed that the temperature dropped by 7°F to 14°F (4°C to 8°C) over the entire American continent, and by 2.7°F to 6°F (1.5°C to 3°C) in Europe. This sudden cold snap lasted three centuries, after which the situation returned to normal.

In 1999, Don Barber and a team of researchers at the University of Colorado finally discovered the origin of this phenomenon. The presence of red sediments from ancient lakes in the Hudson Strait and Greenland, along with mollusks and freshwater plankton, all dating back eighty-two hundred years, confirmed their theory that two huge glacial lakes had suddenly spilled massive amounts of water into the Atlantic.

These lakes had formed in the north of Quebec at the end of the last ice age, near the Hudson Bay. An ice barrier held the water back for years, but then it began to melt. As the barrier broke, tremendous amounts of fresh water poured into the Labrador Sea, modifying local water currents, as well as the ocean's rate of evaporation throughout the entire region. Brutal consequences followed; a cold, dry climate settled over Europe and America for the next three centuries.

This discovery is particularly chilling in light of our current situation. We are no longer faced with two lakes spilling into the North Atlantic, but all the ice in the Arctic Polar Circle!

*"Small changes affecting polar ice could have major consequences for the water cycle and, eventually, the climate."*

—NASA

The ICES satellite launched by NASA recently observed that the glacial ocean has been losing 10 percent of its layer of permanent ice each decade since 1980. (Glacial ocean ice becomes thicker in the winter and partially melts in summer. Ice that never melts is known as "permanent ice.") The amount *of permanent ice reached its lowest level in 2002 and 2003, according to NASA researcher Josefino Comiso.*

*The temperature of the oceans and of land around the Arctic has risen by 2°F (1°C) over the last ten years.*

## POSSIBLE ... OR NOT LIKELY?

We should not discount positions that run counter to these theories. According to researchers such as Gilles Delaygue (of the University of Chicago), most models show that temperature variations over the next few years will be limited to just a few degrees, and will not reach the variations necessary to provoke major climate changes.

As for the melting of the glaciers, researchers such as Robert Vivian (ex-director of the CNRS Alpine lab in Grenoble) recently defended a position that contradicts the observations made in this chapter. According to Vivian and like-minded researchers, "only a small number of residual glaciers, situated on the edges of the glaciation zones, are shrinking." These same researchers have criticized their colleagues for "most frequently focusing on glacier tongues, which are easily accessible, and having limited knowledge of the main part of the glacier, which is situated at high altitudes." Others have held that the second half of the twentieth century was characterized by a stabilization, or even an extension of the glaciers, rather than shrinking.

Thus, scientists are not unanimous in their opinions, and it is important to remain open to these kinds of criticisms. Which positions are right? And what are the latest observations from the field?

Unfortunately, the current data seems to prove the most pessimistic thinkers right. Indian researcher and IPCC head Rajendra Pachauri recently declared that "glaciers in the Himalayas are shrinking at an alarming rate." The same observations have been made in the Andes. The layer of ice that peaks at 17,390 feet (5,300 meters) near La Paz, on the Chacaltaya, has even split in two, according to researcher Edson Ramirez. The same observations have been made on nearly every one of the planet's peaks.

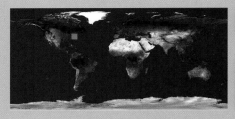 

Ice floes surround the island of Manhattan, in New York City. The Gulf Stream has disappeared, ending the temperate climate previously enjoyed by North America and Europe.

*"It is happening right now. We cannot afford to wait a long time to find technological solutions."*

—David Rind (NASA Goddard Institute for Space Studies), press conference, Washington D.C., 2003

*"By trying to make life simpler, man only succeeds in making it far more complicated."*

—Proverb

The Cathedral of Basil the Blessed, or Saint Basil, on Red Square pokes through the thick coat of snow that covers most of Russia.

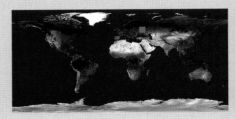

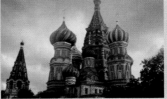

"*First and foremost, by protecting nature, man protects man: he satisfies the species' conservation instinct.*"

—**Jean Rostand**, in the foreword to Édouard Bonnefous, *L'homme ou la nature? [Man or Nature?]*, 1970

London under the ice. Here, the Thames is frozen solid. It no longer flows under Tower Bridge, which is trapped under ice.

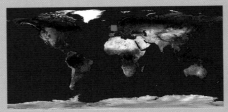

The Atomium in Brussels, Belgium, the only monument rising from the snow.

*"To surmount an obstacle, you have to look beyond it."*

—Proverb

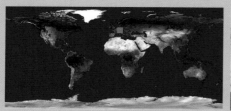

Also in Europe—the city of Munich, Germany.

*"The most important step toward solving a problem is to recognize that you are part of the problem."*

—Proverb

The Chateau de Chambord in France, gradually disappearing under the snow.

"The fate of the world will be such as the world deserves."

—**Albert Einstein**, *The World as I See It*, 1934

The Mont-Saint-Michel sits in the heart of an iceberg field stretching all around England and the French shores of the Channel.

"To resolve a problem wisely, you must turn your attention to it."

—Proverb

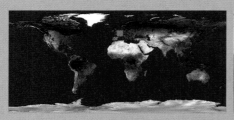

Glaciation underway. The interior of France gradually gets covered over with a thick layer of snow, as can be seen here in Metz, in Lorraine.

"By allowing man to exist, nature committed more than an oversight:
it allowed for a brutal attack on itself."

—Émile Cioran, *De l'inconvénient d'être né*
(On the Inconvenience of Being Born), 1973

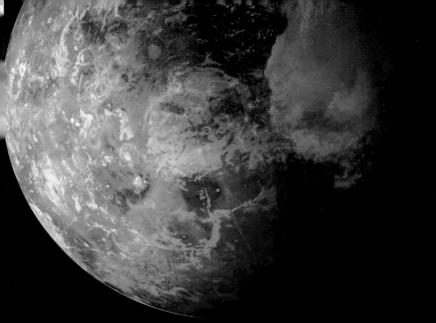

# THE SIXTH EXTINCTION

I N this chapter, we will consider the future of the earth ... without humans. Our objective is to put man in his place and to develop a clear-sighted view of the role mankind has played on Earth.

## HOMO SAPIENS, JUST ANOTHER SPECIES

Our species only appeared on this planet a few hundred thousand years ago. That's just a blip in time if you consider man's reign in comparison to that of the ant or the crocodile.

Man is the fruit of a very long, 3.8-billion-year evolution. Of the numerous species that have walked the earth, 99 percent are now extinct. The extinction of *Homo sapiens* would therefore be far from exceptional.

Nuclear wars, skyrocketing pollution, increasingly prevalent deadly diseases (cancers, respiratory illnesses, etc.), and a whole slew of natural disasters are among the many threats already hanging over us. The human victims of the degradation of our environment are countless. Though many people find the downfall of human societies inconceivable, it remains a distinct possibility.

Our current civilization would not be the first to succumb. Remember that hundreds of civilizations have collapsed throughout the history of humanity. Using our power and technological advances as arguments to prove that our society can withstand any natural aggression is pure folly. Technology may uproot us from our natural environment, but it does not make us independent from it. In fact, didn't technology once come close to wiping us off the face of the planet? Think back to the Cuban missile crisis....

## THE SIXTH GREAT EXTINCTION

Paleontology, which tracks the evolution of life using fossils, has taught us that life on Earth has gone through five major crises (Appendix 4). The most recent and most famous crisis took place 65 million years ago and caused the disappearance of the dominant terrestrial species, the dinosaur reptiles, despite their 165 million years of evolutionary experience.

Though the five previous crises were entirely due to natural disasters (meteorites, volcanic eruptions, etc.), the one we are experiencing today is the first caused by a species' activities. Could intelligence be a factor of extinction?

Though the origins of the crises are still hotly debated, one fact is indisputable: Each crisis resulted in a brutal modification of the planet's environment. As incredible as it may seem, the situation we are currently facing is entirely comparable to these five crises, to the point that our era has already been designated the "Sixth Extinction."

Consider that 30 percent of current species could be extinguished by 2050, and that the current rate of extinction of species is one thousand times greater than before the industrial era.

An average of between 1 and 10 percent of species are eliminated per decade. At this rate 1 million species will disappear within twenty-five years! As for plant life, 32 million wooded acres vanish per year (the earth has a total of nearly 10 billion acres). Deforestation remains catastrophic in South America and Africa (causing the quantity of carbon captured by forests—which currently stock about 283 gigatons of carbon—to diminish by an average of 1.1 gigatons per year since the beginning of the 1990s).

Today, 11 percent of mammals, 13 percent of fish, 10 percent of amphibians, 8 percent of reptiles, and 4 percent of birds are endangered. These figures match those for the previous five crises.

It goes without saying that any reference to the term "mass extinction" is far from arbitrary.

FORESTS:
More than half of global forests were destroyed during the twentieth century; 80 percent of the varieties of trees have disappeared. This deforestation is accelerating at a rate of two football fields per second! (Fifteen percent of the Amazon rain forest was razed between 1970 and 2000.)

SEAS:
In the North Sea, mackerel supplies have collapsed; the salmon population is getting dangerously low; the ray has nearly vanished (due to troll nets); the haddock, whiting, herring, and

cod are threatened; the drop in whiting has reduced the number
of Steller sea lions off Alaska by 90 percent; and the number of
Japanese sardines has dropped by 90 percent in twelve years.
In Canada, 27 percent of belugas show signs of cancer (only six
hundred fifty remain), and the number of Atlantic salmon has
dropped by 70 percent over the last twenty years. In 1998, the
extended warming of the water destroyed half the coral reefs
in the Indian Ocean.

## MAN THREATENED WITH EXTINCTION

By a cruel twist of fate, mankind, which has long been mauling its environment, may very well be in the
process of causing its own disappearance. Countless threats already hang over the species, and the effects of
pollution have already caused millions of victims. A quick overview of the threats:

### THE PESTICIDE THREAT

Used abundantly by farmers, pesticides destroy vast areas of land and poison humans, leading us to
develop a variety of illnesses. In 1980, the World Health Organization (WHO) confirmed that pesticides
caused twenty thousand deaths and poisoned 3 million people a year—figures that have since grown
considerably.

The constant increase in genital and breast cancers are allegedly directly linked to these contaminations. (In
1950, one in twenty Americans developed this type of cancer; today the figure is one in eight.) Additionally,
polychlorinated biphenyls (PCBs) cause many hormonal abnormalities among males. Sperm rates have
decreased by 50 percent since 1945.

In France, 10 percent of farmland is too polluted for cultivation.

Farmers pour fertilizers on their land in abundant amounts,
yet fertilizers are no less dangerous. Only 10 percent of these

*products stick to plants; the remaining 90 percent get disseminated in nature as extremely toxic chemical compounds (e.g., nitric acids, nitrogen oxide, and nitrates) that pollute water sources.*

## THE HOUSEHOLD PRODUCT THREAT

Found in every home, household products are among the most carcinogenic: Nearly all glues, varnishes, and cleaning sprays contain highly dangerous compounds, such as formaldehyde, which is commonly used to polish furniture.

*Four hundred million tons of these substances were produced in 2003—1 million tons were produced in 1930.*

*"As a cancer specialist, I have noticed that cancer is a disease entirely produced by our society and largely due to environmental pollution.... The conclusion to be reached is obvious. Today's diseases are not the natural diseases of the past. They are all, or nearly all, artificial."*
—Professor Dominique Belpomme

*According to cancer specialists, more than 90 percent of cancers (which currently affect nearly every family) are due to environmental pollution.*

## THE GROWING, GENUINE THREAT OF GMOS

Though agroindustrial lobbies continue to impose genetically modified organisms (GMOs) on us, it has been clearly demonstrated that these products are a menace. Numerous tests have revealed the development of serious diseases in laboratory animals. The United Nations Report on the Right to Food is very clear on the subject:

*"GMOs can pose dangers in the medium and long term for human beings and for public health."*
—U.N. Special Report on the Right to Food, November 2002

### THE DOUBLE OZONE THREAT

There are two types of ozone. The good type is present in the stratosphere and protects us from UV rays. Necessary for life, this type of ozone has been watched very closely since the detection of large "holes" in the stratospheric strata over the last few decades.

The shrinking of the ozone layer could lead to a massive increase in skin cancers, for the decrease in chlorofluorocarbons (CFCs, the leading cause of stratospheric ozone depletion over the last few decades) has been compensated by an increase in methyl bromide, a pesticide forty times more dangerous for ozone molecules than CFCs and commonly used in developing countries.

A second type of ozone—bad ozone—is produced by automobiles. Its potently dangerous oxidizing power modifies the permeability of our cellular membranes. Contrary to good ozone, bad ozone is a molecule we'd love to eradicate.

### THE CLIMATE BARRIER DISPLACEMENT THREAT

Over the last several years, experts have observed a displacement of climate barriers due to global warming. This displacement has led to a migration of certain species, including an ominous northern migration of many mosquito populations.

Mosquitoes, the principal carriers of serious diseases, now threaten to cause severe epidemics and pandemics among the populations of Europe. In fact, several animal species have already fallen victim to this trend. Catharral fever, which was carried by insects migrating from Africa to Corsica by way of Sardinia, decimated hundreds of Corsican sheep. Likewise, horses in the Camargue have been suffering from West Nile virus for several years. Similar insect migrations could propagate diseases such as malaria throughout Western nations.

## A FEW HARD NUMBERS

Here are a few hard numbers referring to the population of France (these figures can be even more alarming in other developed countries).

*In 2004:*
*33% of male deaths were from cancer*
*25% of female deaths were from cancer*

*Between 1980 and 2000 there was:*
*63% rise in cancer cases (figures proportional to industrialization)*
*30–50% increase in child leukemia, lymphoma, and brain tumors*
*200% rise in birth defects*

*Today:*
*15% of couples are sterile*
*280,000 additional cancer patients each year*
*6,500–9,500 deaths annually attributed to air pollution*
*20% of the population suffers from allergies (2,000 deaths a year)*

## MAN IS ON THE WAY OUT, BUT LIFE GOES ON

The extinction of humanity, should it come to pass, would not necessarily imply the disappearance of life. As any biologist would be quick to tell you, the living model is astonishing not only for its complexity and ingenuity, but for its robustness. It is more than likely that bacteria or microscopic extremophilic animals would survive even a nuclear winter.

The Antarctic ozone hole (seen here in 2000). Stratospheric ozone protects us from the sun's ultraviolet rays. The disappearance of this shield due to emissions of industrial gases such as methyl chloride threatens to cause a major increase in skin cancers and ocular lesions in the coming years.

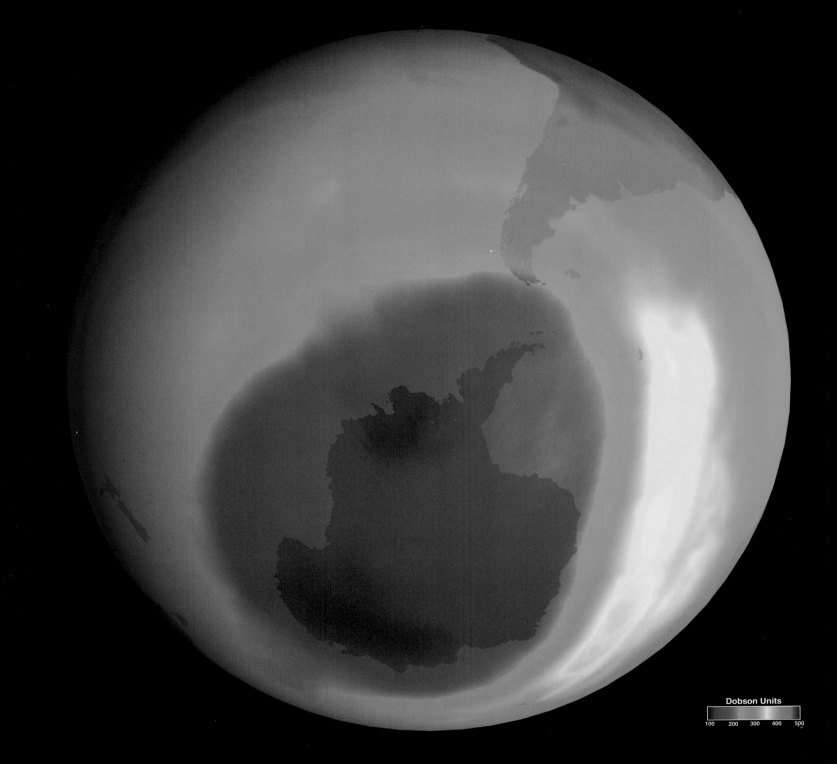

Dobson Units

100 200 300 400 500

What impact would man's extinction have on the environment? Probably an adaptive radiation.

For example, let's examine the Cretaceous Period, which ended 65 million years ago. During the Mesozoic Era, dinosaurs reigned over terrestrial ecosystems. Upon their extinction, many ecological niches were left vacant, and many species, including our ancestors the mammals, were left to benefit from this change. Each extinction has been followed by an adaptive radiation: an outburst of life.

The disappearance of the human race would probably be followed by a similar global phenomenon, for man's presence is currently felt in every ecosystem, both on land and in the water. As for climate, new equilibriums will form, and a return to a stable situation should happen quite quickly (albeit on a scale of a few centuries). For example, $CO_2$ rates were up to ten times greater than today during periods such as the Mesozoic Era, when the earth's temperature was nearly 20°F (10°C) higher than it is today. Though this global warming could decimate humanity by destabilizing our civilizations, it is highly unlikely that it would be fatal to all terrestrial species.

The pictures on the following pages depict the above scenario. The next chapter will consider a final scenario in which all species are eradicated due to a specific problem: a potential uncontrollable acceleration of the greenhouse effect.

Sixty feet (18 meters) long and weighing nearly 15 tons, a patagosaurus walks along waterfalls in the area that will become Argentina in some 163 million years. The dinosaurs' dominance over terrestrial ecosystems for 165 million years did not save them from succumbing to global environmental change 65 million years ago. Will man be able to survive an equally vast biological crisis?

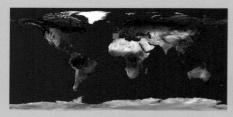 

Dense vegetation has conquered Madrid. By causing the extinction of the human race, global warming has allowed nature to reclaim its place as it creeps over the pavement and erodes ancient monuments built centuries ago by a highly evolved species of primates.

"Every phenomenon man is involved with takes place at an accelerated speed and a rhythm that makes it nearly uncontrollable."

—**Jean Dorst**, *La nature dénaturée* (Nature Denatured), 1965

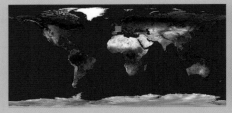

Morning dew evaporates over the vegetation suffocating the Louvre Museum in Paris. On the main square, only the central glass pyramid rises out of the greenery.

"*Nature does not play favorites, and man, despite his genius, is not worth any more to nature than the millions of other species produced by life on Earth.*"

**—Jean Rostand,**
*Pensées d'un biologiste* [Thoughts of a Biologist], 1973

"*Nature does not endure sudden mutations without tremendous violence.*"

—**François Rabelais**, *Gargantua*, 1534

Capitol Hill, in Washington, D.C., covered in dense vegetation. The shells of tanks and traces of combat lead one to believe that a conflict eventually raged deep into the heart of the United States, to its very capital.

"When a species fails to find its purpose, it dies, or destroys itself."

—Satprem,
*La Révolte de la Terre* [The Revolt of the Earth], 1990

Thick vegetation chokes the avenues of downtown Toronto and gradually covers the faces of the city's highest skyscrapers.

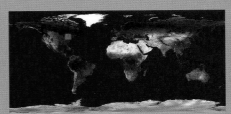

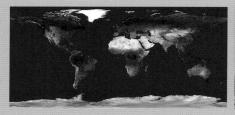

The Acropolis in Athens. A tropical rain forest, such as those currently found in Brazil, covers all the Mediterranean countries, transforming ancient landscapes.

*"Nature: a loan from God to man."*

—*The Koran*

Creepers devour the Hollywood sign overlooking Los Angeles. Lush forests have replaced the arid land of California.

"We are predators full of remorse and scruples: man had barely appeared in the biosphere when he started wondering about the danger he posed to the planet sheltering him.... In fact, it's not the planet that's in danger, it's man. Blinded by his unrivaled power, man cannot see that he is sawing off the branch on which he sits flexing his brainpower."

—Jean-Louis Étienne,
*Le Pôle intérieur* [The Interior Pole], 1999

"We are engaged in a huge climate experiment on a global scale. We are observing those effects that are already clearly visible, and anxiously watching out for those to come."

—**Hubert Reeves**, *Mal de Terre* [Earth Sickness], 2003

The ruins of the famous Millau Viaduct in France. Part of the platform has collapsed, but the celebrated silhouette is still recognizable over the Tarn Valley.

The heart of New York City, overrun with vegetation. Here, Times Square is disappearing under a tangled layer of greenery.

"Man must be nature's guardian, not its owner."

—Philippe Saint-Marc, *Socialisation de la nature*
[Socialization of Nature], 1971

"*The violence found in nature must not serve as a justification for man's violence. On the contrary, this reality provides man, as a moral being, with an additional duty.*"

—**Théodore Monod**, *Terre et Ciel* [Earth and Sky], 1999

The skyscrapers of downtown Singapore are devoured by a lush forest. The high towers, which have been transformed into nesting areas for the many bird species in the region, are gradually becoming decrepit, yet they appear to continue to hold off the onslaught of vegetation.

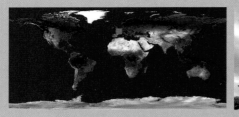

The rise in global sea levels led to the flooding of the Paris Basin, which had long since been abandoned by humans. Global warming caused a hot and humid climate similar to the one found in this area tens of millions of years earlier.

"Man's stubborn behavior in the face of nature conditions the stubborn behavior among men."

—**Karl Marx**, *The German Ideology*, 1845–1846

*"When the last tree is felled,
The last river poisoned,
The last fish is fished,
Then you will discover
You cannot eat money."*

—Cree saying (Canadian Indians)

The extinction of mankind would probably not imply the end of life on Earth. Other species would fill the ecological niche left vacant by *Homo sapiens*, including (why not?) other forms of advanced intelligence.

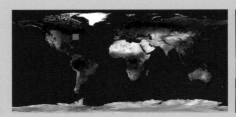

LAND OF FIRE

THE worst-case scenario involves a sudden, uncontrollable acceleration of the greenhouse effect. In his book *Mal de Terre* [Earth Sickness], astrophysicist Hubert Reeves imagines the consequences of such a phenomenon. Three scenarios are developed based on specific temperature rises. The first would see a global temperature rise of a few degrees Celsius by 2100. Reeves refers to this scenario as "Desert Possibility." It corresponds to the potential futures described in the previous chapters. Unfortunately, Reeves is here to remind us that a sudden acceleration of the greenhouse effect could make global temperatures rise far more dramatically, leading to two other potential scenarios.

The "Geyser Scenario" describes the effect of a global temperature rise of several dozen degrees Celsius.

According to this scenario, all multicellular organisms, as well as insects, would disappear from the planet's surface. Only bacterial life would remain. Indeed, bacterial organisms have already demonstrated their ability to survive in extreme situations, such as those found in Icelandic lakes that expel boiling water through geysers. If this scenario were to take place, the evolution of life would move backward approximately one billion years to return to forms of life identical to the very first organisms that developed on Earth. From *Mal de Terre:*

> "Human aggression will have succeeded in doing what to our knowledge no other geological or astronomical activity has succeeded in doing: returning life to its bacterial, single-celled form."

Finally, he writes about the "Venus Scenario," where temperatures could increase even more, spiking over 212°F (100°C). Though unlikely, this phenomenon is not impossible, particularly since we have a literal example of another planet in the solar system that displays those very conditions: Venus.

The quantity of $CO_2$, the principal greenhouse gas in our atmosphere, has increased rapidly since the beginning of the industrial era. Amounts of atmospheric $CO_2$ have increased by a third since the beginning of the twentieth century. Its concentration is currently the highest in six hundred fifty thousand years. It is expected to double within a century.

A photo of Venus taken from the Galileo Probe (NASA).

## THE GREENHOUSE EFFECT EXPLAINED

Imagine a greenhouse containing compost. Sunlight passes through the greenhouse's window and is absorbed by the compost. As the compost's temperature increases, it releases heat in the form of infrared rays. However, the window does not allow infrared through and the heat does not evacuate. The temperature in the greenhouse increases.

On Earth, water vapor, methane, and carbon dioxide act like the greenhouse window. Though dangerously accentuated in recent years by human activities, the greenhouse effect is actually a natural phenomenon necessary for the planet and the existence of life. Without this phenomenon, the average temperature on the surface of the earth would be −0.4°F (−18°C), versus 59°F (15°C) under existing conditions, and very little water would be found in liquid form.

## VENUS, THE "GREENHOUSE" PLANET

Scientists consider Venus to be the Earth's twin. The two planets share many essential characteristics: They have similar masses, are about the same distance from the sun, and contain practically the same amount of carbon.

The major difference resides in the amount of carbon dioxide present in the two planets' atmospheres; 96 percent of Venus's atmosphere consists of carbon dioxide, which creates an important greenhouse effect. As a result, the temperature on Venus is over 860°F (460°C), which is hot enough to make lead melt and produce sulfuric acid rains over the planet*. No form of life as we know it could survive on Venus.

*The data gathered by the Venera Probe revealed temperatures on Venus of 890°F (477°C).

Why has Earth avoided the same fate? Because of water, the liquid form of which is found in abundance on the surface of our planet. Most $CO_2$ is found in the oceans, where plankton have transformed it into sediments. On Venus, there is no water. However, while levels of $CO_2$ in our atmosphere previously remained constant, the quantity of this gas has now increased to the point that the oceans can no longer fulfill their role as buffers. From *Mal de Terre:*

"The Venus scenario would be a four-billion-year step backward in the development of cosmic complexity on our planet. The earth would return to the state it was in in its first hundreds of millions of years, before a layer of water appeared, before the first molecular reactions that triggered the extraordinary and still largely mysterious process by which terrestrial life appeared, before the first living cells that reigned over the planet for more than three billion years."

## THE CURRENT SITUATION

Meteorology provides us with data about recent climate conditions (as of the 1800s); in order to go back several centuries, we study trees or corals. Paleoclimatologists study the proportions of isotope-18 in oxygen (which increase with temperature) held in Antarctic ice for periods dating back several millennia to several hundred millennia (up to nine hundred thousand years). Today, studying stalagmites or stalactites in caves allows us to go back several million years.

> Since we began recording weather conditions, in 1861, the average temperature has risen from 31.28°F to 33.44°F [-0.4°C to 0.8°C] for the entire globe.

> The average temperature in Western Europe has already risen to 33.8°F[1°C] in one hundred years. Since the 1980s, the rate has rapidly accelerated.

The volume of $CO_2$ in our atmosphere is currently 380 ppm (parts per million). It was only 280 ppm during the preindustrial period. (It should be noted that human activities release seven gigatons of carbon into the atmosphere per year, one gigaton due to deforestation, and six gigatons due to hydrocarbons.) Analysis of new ice cores in Antarctica (through the EPICA Project—European Project for Ice Coring in Antarctica) have recently (early 2006) confirmed that concentrations of $CO_2$ and methane currently measured in the atmosphere are the highest recorded in more than six hundred fifty thousand years. (These cores, which at 10,500 feet/3,200 meters are the deepest ever drilled, have allowed researchers to extend their data another 210,000 years—two complete additional glacial cycles beyond the Vostok core.)

## DETAILS OF A DOOMSDAY SCENARIO

The rise in global temperature, which will initially melt ice and cause sea levels to rise, would be followed by the evaporation of ocean water, which would contribute to further increasing temperatures. The Canadian and Siberian permafrost would melt entirely, provoking the release of massive amounts of methane held in the ice's crystalline lattices*.

*See also "The methane hydrate threat," p. 137.

Methane is a greenhouse gas several dozen times more efficient at warming the atmosphere than $CO_2$ (in equal concentrations, methane contributes to the greenhouse effect seventy times more than $CO_2$). There are 390 gigatons of methane trapped in the crystalline lattices of the Northern Hemisphere's ice banks.

Once again, this phenomenon could suddenly accelerate, exposing the surface of the earth to extremely high temperatures that would progressively rival those on Venus.

Note that the melting of the Siberian and Canadian permafrost is already under way, and that the release of methane into our atmosphere has already begun—to a point where growing numbers of American farmers are buying land in Alaska in hopes of growing wheat there in the coming years.

According to a recent study undertaken as part of the "Climate-prediction" project, the coming temperature increase will be between 4°F and 20°F (2°C and 11°C), a much greater range than the one predicted by the U.N. (between 4°F and 9°F/2°C and 5°C). This study, launched in 2003, was modeled on the seti@home project to seek out extraterrestrial intelligence. The project uses ninety thousand personal computers spread through one hundred and fifty countries according to the principle of grid computing.

*The climate warming predicted in these measurements will be exceptional: It is the most rapid temperature change the earth has ever experienced. Over a few centuries, we will experience changes that would previously have stretched over millennia, including during the most catastrophic periods in the earth's history.*

## THE METHANE HYDRATE THREAT

We have just seen that global warming can melt the permafrost and release enormous quantities of methane, a veritable monster hibernating in the crystalline lattices of polar ice.

Recent studies of the Permian Crisis (the most significant crisis faced so far by life on Earth) have uncovered the existence of another, even more dangerous monster lurking in our oceans.

For many years, paleontologists attributed the greatest crisis in the history of life on Earth (during which more than 90 percent of species were exterminated) to a potential collision with meteorites. Yet no significant evidence has ever been found to support this theory.

Evidence of the Cretaceous Crisis, for instance, is characterized by the presence of a layer of iridium (a metal rarely found on the surface of the earth, but found in great abundance in space) separating Mesozoic and tertiary sedimentary strata. Upper Cretaceous strata display other clues such as traces of massive tsunamis, impacted quartz crystals (requiring extreme pressure to form), or simply impact craters. None of these clues has been found in the higher strata of the Permian.

Finally, in the late 1990s, Paul Wignall, a geologist at the University of Leeds, discovered the beginnings of an explanation in the Permian strata in Greenland.

He initially demonstrated that this extinction spread over tens of thousands of years (in fact, three long phases that followed one another over eighty thousand years), contradicting the hypothesis of a sudden catastrophe such as a meteorite collision. Next, Wignall became intrigued by the abnormally high quantity of carbon-12 in his samples.

Carbon-12 is a particular form of carbon normally produced by the decomposition of organic matter. However, the proportion found in the samples of Permian rock was much too high to be explained solely by this phenomenon.

The solution to the problem came during a barroom discussion, where Gerald Dickens, a geologist at Rice University in Texas and a friend, wanted to know how large quantities of carbon-12 could be produced naturally and quickly.

Some years earlier, Dickens had worked on drilling for a new source of energy in the Gulf of Mexico: methane hydrates. Gas hydrates (also known as clathrates) only form under very specific pressure and temperature conditions, in which ice can trap gas molecules such as carbon dioxide, hydrogen sulfide, or methane (resulting in methane hydrates) in its crystalline lattices. As for methane gas, it is due to the degradation of organic matter by anaerobic bacteria at the bottom of the ocean.

Methane hydrate is a crystal containing an enormous quantity of gas (the melting of 0.06 of a cubic inch/1 cubic centimeter of methane hydrate releases 10 cubic inches/164 cubic centimeters of methane). Large deposits of these hydrates are found along coasts, on the earth's continental plates (the only regions with the pressure and temperature conditions required to form them). Dickens was aware that these hydrates also contain large amounts of carbon-12.

Could these hydrates have been the solution to the mystery of the Permian Crisis? If so, we needed to understand how the methane had moved from the depths of the ocean to the atmosphere.

Dickens was quick to find the solution. Hydrates placed in slightly warmed water (41°F/5°C was sufficient) break up and release a very large amount of carbon-12-rich gas.

Upon hearing of Dickens's work, Wignall realized the terrible chain of events that led to the extinction of nearly every terrestrial species 225 million years before man.

At that time, the Siberian Traps, a large igneous province, erupted, releasing massive quantities of gas into the atmosphere and increasing the greenhouse effect. Models show that this eruption increased the earth's temperature by approximately 9°F (5°C).

Though this temperature increase would result in certain climate changes and the attendant extinction of several species, it would not be sufficient to extinguish 90 percent of the species on Earth. However, it was sufficient to cause the breakup of all of the planet's methane hydrate deposits. Gases released into the atmosphere by hydrates increased global temperatures by another 9°F (5°C), at which point the total global temperature rise reached 50°F (28°C), enough to exterminate most living species.

The link between this cataclysmic episode and our own situation is obvious and chilling.

Humans are currently playing the part that the Siberian Traps played 225 million years ago. Every model predicts a significant rise in global temperatures within a few decades, an increase that could potentially be greater than the one caused by the Siberian Traps during the Permian, or more than enough to rouse that titanic sea monster dwelling at the bottom of the ocean. Our own impact on the environment, which could single-handedly cause vast catastrophes in the coming decades and centuries, may wind up being little more than the spark that sets off a succession of apocalyptic phenomena.

## REDUCING OUR USE OF FOSSIL FUELS

The one and only key to limiting global warming would be to immediately halt our consumption of fossil fuels (principally oil).

Just as certain ecologists had begun celebrating the depletion of natural oil reserves, the melting of the Glacial Ocean is giving industrialists hope that new fossil fuel resources will be discovered in the North Seas ... and President George W. Bush, who persists in refusing to sign the Kyoto Protocol, has already launched a study to exploit oil in Alaska in the Arctic National Wildlife Refuge, despite the fact that this is a protected area. Although the United States remains the greatest polluter in the world, it refuses to accept any initiative out of fear of *changing the American way of life*. Russia's recent *in extremis* signing of the Kyoto Protocol has further isolated the United States. (Or, rather, *some* in the United States. Most citizens do not share the opinions of the federal government on this issue—the public is mounting increasing pressure on the current administration for its lack of leadership and initiative in the field of environmental protection.)

*The Kyoto Protocol, signed in 1997, only called for a 5 percent reduction in the 1990 emissions of the thirty-eight richest countries on the planet by 2012. As English academic Martin Parry, who was responsible for a study on the consequences of rising temperatures in Europe, noted: "Keeping to this reduction rate would mean reducing a 2° Celsius [4°F] temperature rise in 2050 by 0.06° Celsius [0.10°F]." Source: Libération, Sylvestre Huet, February 22, 2001.*

## OUR HISTORICAL RESPONSIBILITY

If such a cataclysmic scenario comes to pass (let us insist on "if"), we would be responsible for the greatest of tragedies: the disappearance of life. Our actions are already linked to the disappearance of many species, an unprecedented occurrence in the history of evolution. Predator and prey have always existed, but the human race has invented a new category: the eradicator.

By playing this dangerous game, we could very well wind up being our own executioners, for without those other species we are fated to disappear. Our environment is symbiotic. Though a stubborn few continue to cling to the supremacy of man by hijacking religious beliefs, the truth couldn't be more different: Man is a species among others. This planet does not belong to us. We are only its tenants, and certainly not its owners.

Some may accuse me of being overly dramatic for considering such a distant possibility. Let me reiterate that this book only deals in possibilities, not dogmatic affirmations. However, paleontological studies have revealed that similar scenarios have taken place in the past; astronomy has allowed us "live coverage" of similar phenomena (such as the greenhouse effect) and their effects on planets such as Venus; and the latest research in climatology demonstrates that the conditions necessary to unleash these destabilizing events are close to being in place.

*Rampant desertification in certain areas of the globe has forced man to abandon once-fertile areas such as this part of Yemen.*

One of the biggest ships in the world, *Queen Mary 2*, washes up on what was once the bottom of the sea.

*"Forests precede the people, deserts follow them."*

—**François-René de Chateaubriand**, *Itinéraire de Paris à Jérusalem* (Itinerary from Paris to Jerusalem), 1811

"Earth managed to get by without man for many millions of years: he was, let us not forget, the last living creature to appear on Earth. So, either he will accept to change his view of the world, and his philosophy, or we will see other groups replace him."

—**Théodore Monod**, *Terre et Ciel* [Earth and Sky], 1999

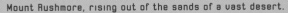

Mount Rushmore, rising out of the sands of a vast desert.

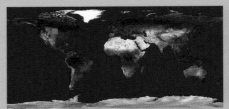

"*If all the animals disappeared, man would die, for what happens to animals soon happens to man.*"

—Seattle, Indian chief, from a declaration to the president of
the United States, 1854

The White House stands abandoned. Like ancient Rome, the once-mighty American superpower is reduced to a stack of ruins eroded by the arid sand.

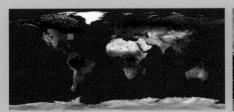

The Venice Lagoon has dried up, uncovering the sea bottom. Only a few corroded wrecks serve as reminders of what was once a sea.

"Man appeared like a worm in a fruit, like a mite in a ball of wool, and gnawed at his habitat, secreting theories to justify his actions."

—**Jean Dorst**, *La Nature dénaturée* (Denatured Nature), 1965

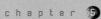

"*I am never sure. Expect the unexpected, I think the future is unpredictable, the worst can happen.*"

—**Jean Baudrillard and Edgar Morin,**
*La Violence du monde* [The Violence of the World], 2003

A sandstorm over the ruins of Kuala Lumpur.

*"Perhaps we're condemned to seeing we're headed for disaster, without being able to turn back?"*

—Robert Kandel,

*Le Devenir des climats* (The Future of Climate), 1995

The Montreal Olympic Stadium is one of the rare buildings in the city not to have been entirely buried by sand.

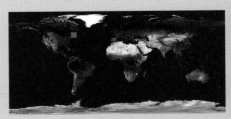

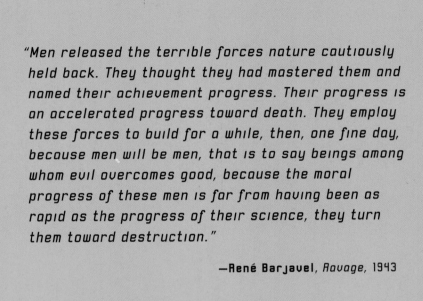

"Men released the terrible forces nature cautiously held back. They thought they had mastered them and named their achievement progress. Their progress is an accelerated progress toward death. They employ these forces to build for a while, then, one fine day, because men will be men, that is to say beings among whom evil overcomes good, because the moral progress of these men is far from having been as rapid as the progress of their science, they turn them toward destruction."

—René Barjavel, *Ravage*, 1943

The Great Wall of China no longer divides anything but fields of dunes stretching as far as the eye can see. The Gobi Desert grew to the point that it covered all of Asia.

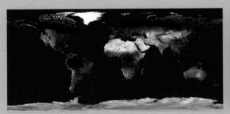

The dried-out Bay of Hong Kong. Scattered detritus testifies to the regular passage of groups of nomads.

"When the well is dry, you know the value of water."

—Proverb

# THE HUMAN REVOLUTION

WE'LL end with a chapter full of hope, the hope that humanity quickly becomes aware of the danger it faces. The next few pages will describe a scenario completely different from those found in the previous chapters: With their back to the wall, humans realize their errors and completely modify their society and their relationship to nature. For solutions do exist. Without rejecting all technology and returning to the Stone Age, it is still possible to transform our societies so a harmonious relationship exists between *Homo sapiens* and their environment.

Together we'll take a look at the various solutions we have available and imagine what we could make of the world.

## A NECESSARY AWARENESS OF THE RISKS WE FACE

It's hard to imagine that one day our societies could collapse and live on only as ruins. Scientists today are faced with the difficult problem of making the general public aware that today's fictions could become tomorrow's realities.

Yet similar catastrophes have taken place. On a smaller scale, natural disasters have caused the downfall of genuine civilizations, some of which had rich histories stretching back thousands of years.

The ruins of Machu Picchu in the Andes (left) and the Great Pyramid and the Sphinx on the Giza Plateau (below).

What reactions might members of these civilizations have had if we had been able to show them tourists' snapshots of their gigantic pyramids worn away by erosion, their temples buried by dunes, their towns submerged in water, or the remains of their cities suffocated by dense vegetation? Nonetheless, these are the landscapes we see when we look at the pyramids on the Giza Plateau, the temples of the Valley of the Nile, or the pre-Columbian towns of Palenque or Machu Picchu, to mention some of the most famous examples.

We must realize how fragile we are. The last decades of archeological and paleontological research have allowed us a totally unprecedented depth of perspective. We have learned about the collapse of past civilizations but also of the total extinction of countless living species during major biological crises.

## WHY DO WE NEED TO RETHINK OUR PRIORITIES?

In order to radically change our course, we must first define new priorities.

Certainly, the threat of terrorism should not be taken lightly. Weapons of mass destruction are circulating more or less freely throughout the world; significant quantities of radioactive material have already been stolen; and autocratic governments are still in place. Yet if you step back and consider the damage done in the last few years by climate-related natural disasters, it becomes obvious that most governments underestimate the risks of climate change.

According to the secretary general of the U.N., the cost of climate-related disasters is seven times the cost of wars.

Today, the number of ecological refugees is greater than the number of political refugees.

Where does the sea end and where does the continent begin? It's hard to gauge from this satellite photo of the ruins of New Orleans (flooded by the waters of the Gulf of Mexico following the landfall of Hurricane Katrina). These flooded neighborhoods will probably never be reconstructed.

*The United Nations University estimates that the degradation of terrestrial ecosystems should force nearly 50 million people into exile by 2010. The official creation of an "environmental refugee" status is currently under consideration.*

Global warming does not threaten a few thousand or even hundreds of thousands of individuals, but humanity as a species, along with numerous contemporary animal and plant life forms.

Faced with these kinds of stakes, we do not have any right to dismiss these threats as highly improbable. Even if there is a low probability of some of these events actually taking place, it is our duty, given the risks we're facing, to make the climate problem our top priority, far ahead of our petty quarrels and, more generally, of our conflicts of interest.

## CHANGE REMAINS POSSIBLE: SOME POSITIVE EXAMPLES

*"Our problems often seem unsolvable because the way we approach them does not allow for a solution."*

Changing the world is necessary and entirely possible. I turn for evidence to the rehabilitation of some large North American lakes and forests that had been devastated during the 1980s by regular acid rain. People's awareness of the catastrophe, their understanding of the causes, and the search for solutions gradually taught Americans that installing filters on the smokestacks of all thermoelectric plants could provide viable solutions. These types of measures were progressively adopted, and the problem is now under control. (At least in that part of the world. We must now help poor countries follow suit.)

Though we are frequently led to believe that the use of many chemical products can stimulate our agricultural production, poisoned populations and ruined farmland led the Indonesian government to issue a blanket ban on fifty-seven pesticides in 1986. The quantity of pesticides used diminished by 60 percent, and rice production increased by 25 percent!

Offshore wind-power plant: an eminently clean solution, and aesthetically pleasing, since it stands beyond the onshore horizon line. Wind-power plants could provide many areas around the globe with power. Unfortunately, if it isn't an interior or political lobby stopping their construction, it's fishermen recklessly fighting a reduction in their fishing areas.

As for the problem of deforestation, China has launched an ambitious reforestation program in the last few years; nearly 2.4 million acres of forest are currently replanted annually, in a country that was previously losing 200,000 acres a year.

## FALSE SOLUTIONS AND REAL SOLUTIONS

*"He who wants to succeed finds a way, he who doesn't finds excuses."*

*The exploitation of natural reserves is very poorly distributed; today, 15 percent of humans consume 80 percent of reserves.*

Which solutions could actually save the planet? There are two types of solutions. Those that resolve a few problems in the short term, but simultaneously create new problems, will be referred to as "false solutions." Those that can permanently resolve our environmental and sanitary problems will be referred to as "real solutions." A brief overview of some of these solutions follows.

### THE FALSE SOLUTIONS

**Nuclear power**  At one time it might have seemed like the solution to global warming. Apparently, nuclear energy does not release greenhouse gases. It appears to make countries with adequate technology independent from oil reserves, which are becoming depleted just about everywhere around the planet and are responsible for far too many political tensions between their owners and consumers.

Unfortunately, the reality is far less cheery. Researchers have gradually uncovered that this type of energy does not make the countries using it more independent*, and that it actually exposes them to new risks (e.g., the Chernobyl disaster). A reactor has an average lifespan of fifty years, and it takes about another fifty years for radioactivity levels to drop enough to safely dismantle the reactor. Who can predict if the countries currently building reactors will be able to afford to dismantle them in a century?

And if we're not sure whether the countries building reactors will be able to dismantle them in a hundred years, what should we expect for the management of radioactive waste over the next one hundred thousand years?

The fact is nobody has truly mastered nuclear technology: no state can manage a major accident; the problem of waste treatment seems insurmountable; dismantlement raises many difficult questions, none of which even

*There have been no recent discoveries of genuine new uranium deposits; Europe, for example, only holds 2 percent of global deposits, which merely leads to a shift in the dependency problem it is currently experiencing with oil.

begin to touch on complex issues such as the warming of rivers or the management of plutonium 239 (which remains active for one hundred thousand years!). Finally, though reactors do not release greenhouse gases, emissions from the entire chain of the nuclear power process (exploitation, refining, combustible enrichment, construction, operation, power plant dismantlement, waste conditioning, and stocking) are far from insignificant. On average, a nuclear power plant using uranium-235 releases one-third the amount of gas that a gas plant would release. Though this ratio currently favors nuclear power, it will not always be so. The amount of $CO_2$ emitted during the extraction and treatment of uranium is proportionate to the amount of uranium in the ore. The rarefaction of uranium reserves will require increasing amounts of power to exploit the uranium, causing the nuclear chain to release more $CO_2$.

The 2003 heat wave pointed to the limits of nuclear technology. Unable to cool their reactors (because the temperature of rivers had climbed too high), many nuclear power plants had to shut down, depriving the population of electricity—and air conditioning. (Given the weather predictions for more frequent and more intense heat waves over the coming decades, as described in the previous chapters, this alone is a significant problem.) Worse yet, engineers have discovered that the water from these rivers can freeze, causing the same kinds of problems during the cold season, when demand for heat drives power consumption through the roof.

Nuclear power is rife with dangers, yet incidents in nuclear power plants (throughout the world) are little known or completely unknown. Consult Appendix 5, at the back of the book, for a list of the most serious incidents to occur in the West—incidents far too frequently brushed over in politicians' florid speeches. This is a chilling list and will certainly make you look at the nuclear reactors in your area in a new light. I invite you to reread the list of nuclear accidents in Appendix 5 and recall the equally reassuring speeches authorities have made regarding the nuclear issue.

GMOS    Let us briefly return to biotechnology. Initially, the modification of living organisms' genetic heritage seemed to promise a beautiful, famine-free future that would allow a variety of crops to develop in even the most hostile soil in the world. Unfortunately, as with every other time man tried to "play God," the experiment quickly showed its limitations. Today, GMOs constitute a major environmental and sanitary hazard: In June 2003, researchers with the Scientific Institute of Public Health listed thirteen negative aspects of GMOs. It has now been proven that GMO crops incorporate extremely dangerous genes in food crops and create super-viruses that penetrate the bacteria of our intestinal flora. In short, yet another technology we've handled with the recklessness of a child fooling around with his father's loaded gun.

Luckily, real solutions do exist, in the energy field—with wind and solar power—and in agriculture—with agroecology. Here, then, is a list of a few solutions, in no particular order. Some are simple—all are respectful of our environment. They should all be adopted by our societies as soon as possible.

**In industry:**
- label all products (food, household, etc.) according to their health and environmental impact to help consumers have a better understanding of what they're buying

**Teach people to:**
- initially, limit their energy consumption
- consume intelligently—buy healthful, harmless products, that respect ecosystems and are nonpollutants

**In agriculture:**
- strictly forbid the planting of GMOs
- impose better fresh water use
- replace pesticides with natural palliatives
- produce and consume organic products (thereby contributing to a reduction in prices)
- develop organic farming through grants and incentives
- push for better crop rotation to achieve more efficient use of farmable land
- replant trees in fields bearing crops, such as wheat, in order to improve the yield

**In transportation:**
- develop railroad freight
- develop public transportation
- encourage carpooling
- encourage the use of nonpollutant transportation alternatives (bicycles, etc.)
- develop new nonpolluting motors using clean power (biofuels, electricity, etc.), while simultaneously developing adequate urban and domestic structures (to recharge electric vehicles, etc.)

**In construction:**
- change construction methods to provide better insulation (limit energy loss due to heating, limit noise pollution, etc.)

- incorporate vast green areas in cities—use roofs to plant vegetation that will limit urban pollution (the notorious smog), while reducing local temperatures
- initiate more efficient use of lighting at night in cities and on roads

**In natural environments:**
- replant our forests in order to create new carbon sinks and sequester $CO_2$, re-create humid zones, and fight flooding
- develop ecological parks, protect and redevelop endangered species, limit fishing and hunting zones
- replant in desert areas to limit the expansion of the deserts

**On the political level:**
- empower government agencies responsible for the environment so that they become the most important parts of the administration—along with the departments of education and culture
- denounce disproportionate relationship of resources and income in Africa and India (to whom we must offer our assistance); these countries are in the throes of development, and we must help them find clean power sources from the beginning and cancel their debt

*"The tripartite agreement [scientific, industrial, governmental] is certainly the essential ingredient for saving the planet. Without it, nothing will change."*
—*Mal de Terre* [Earth Sickness], 1999

**On the technological level:**
- start distributing carbon gas produced in thermal power plants directly into the ocean
- develop solar power (panels, towers, orbital panels) and wind power
- develop the use of hydrogen, which, combined with oxygen, produces water
- develop the use of the biomass
- gradually replace coal power plants with new types of structures (such as wind power plants)

*"If you have the feeling that you can't do much about it, try sleeping with a mosquito...and you'll see which one of the two keeps the other from sleeping."*
—The Dalai Lama

*"The flowers of tomorrow are in the seeds of today."*

—Chinese proverb

The Champs-Élysées returned to pedestrians. Intraurban pollution is responsible for increasingly toxic smog above the world's large cities. A project to relieve congestion in the French capital is currently under consideration.

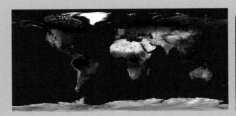

Place de la Concorde covered in gardens.

"Humanity has a latent ability to create something new."

—Jean Baudrillard and Edgar Morin,
*La Violence du monde* [The Violence of the World], 2003

Gardens take over New York City, spreading across rooftops and a variety of neighborhoods. Central Park is no longer the city's only green lung. Smog has disappeared and made it possible to see the heart of Manhattan again. In the distance, new skyscrapers conceived to be wholly environment-friendly tower over the sprawling metropolis.

"One thing is for certain, things must change to improve."

—Proverb

"It takes a lot of courage to escape the bad doctrines that have accumulated around us over the centuries."

—Proverb

Red Square in Moscow. As in many other metropolises, intelligent management of intraurban traffic has made it possible to reclaim vast areas for pedestrians and to develop green spaces in various parts of the capital.

Columbia Hills in the Gusev Crater, Mars. Space exploration reminds us that our planet is like an oasis of life, lost in an immense void. Perhaps one day man will colonize, then "terraform" other planets. Perhaps, unlike his ancestors, he will spread life throughout the universe, instead of destroying everything along his path.

"Every form of life must be considered an essential heritage for humanity. Damaging the ecological equilibrium is therefore a crime against the future."

—Excerpt from "Threats and Promises at the Dawn of the Twenty-first Century," a joint statement by seventy-five Nobel Prize winners, Paris, January 1988

# conclusion

In closing, I'd like to reiterate how important it is for all of us to step back and gain some perspective on the developing environmental crisis. Though some of the disasters described in the first chapter have already taken place, the scenarios in the following five chapters remain mere possibilities, the likeliness of which varies on a case-by-case basis.

Though the goal of this book is to alert the general public to real dangers, it also aims to insist that it is not too late to limit the effects of global warming and to become prepared. Surrendering to boundless pessimism won't do any good. It is time for man to get a grip on himself and to transform his relationship with nature into a symbiotic one. This implies a mutually beneficial connection, with a good "exchange rate." We can no longer afford merely to help ourselves. We must give, protect, respect, and learn.

Nonetheless, the clock is ticking. Climate change is taking place at this very moment. We must react as quickly as possible.

Education and the development of respect for various life forms seem to me to be indispensable steps. We can only hope that future governments will use their departments of education to insist on this type of training. Though most people recognize the crisis at hand, a minority of skeptics (including Michael Crichton, U.S. Senator James Inhofe, and Swede Bjørn Lomborg) continue to distort scientific data to spread doubt, some going so far as to deny the existence of global warming.

Though many uncertainties remain, scientific predictions are based on real observations and extensive prior research. Today, the scientific community no longer questions the existence of global warming. It is our duty to refuse the dark future we are being handed and to build another one. As representatives of the human race, we are all responsible for this difficult task.

Though our nationalities, cultures, religions, and physical and moral characteristics may vary, we are all *Homo sapiens,* a unique species of mammal.

Though our physical appearances, living ecosystems, and diets may vary, we are all beings endowed with life, an exceptional organizational system that appeared on this planet about 3.8 billion years ago.

Since time began, life's viability has depended on a natural balance between living beings and their environment. Paleontology, the study of our ancestors' fossil traces, has taught us that on five occasions this equilibrium was broken.

The first break took place 440 million years ago. Sixty percent of plant and animal species disappeared, including 85 percent of marine species.

The second, 367 million years ago, led to the extinction of 60 percent of species.

The third, 250 million years ago, led to the disappearance of 90 percent of marine species, more than two thirds of reptiles and amphibians and 30 percent of insect species.

The fourth, 208 million years ago, led to the disappearance of 20 percent of reptiles, gastropods, and bivalves.

The fifth and most recent break took place 65 million years ago. Seventy-five percent of species disappeared.

If we brazenly continue along this path, we will probably be faced with a sixth break. These disaster scenarios describe the *future*—a not-so-distant future, but the future nonetheless. So let us not forget that the future is not yet written, and that it remains to be invented.

# appendixes

## APPENDIX 1: LIST OF CYCLONIC EVENTS IN 2004

Sixteen hurricanes and tropical storms devastated the region in less than two months. It began in August 2004, with Hurricane Alex (Category 3) off South Carolina, followed by Charley (Category 4), Danielle (Category 2), Frances (Category 4), Gaston (Category 1), Ivan (Category 5), Tropical Storm Jeanne, which turned into a hurricane (Category 3) as it reached Florida, Karl (Category 4), and Lisa (Category 1).

## APPENDIX 2: LIST OF CYCLONIC EVENTS IN 2005

In order of formation: Cindy (Category 1), Dennis (Category 4), Emily (Category 5), Irene (Category 2), Katrina (Category 5), Maria (Category 3), Nate (Category 1), Ophelia (Category 1), Philippe (Category 1), Rita (Category 5), Stan (Category 1), Vince (Category 1), Wilma (Category 5), Beta (Category 3), Epsilon (Category 1).

## APPENDIX 3: THE SAFFIR-SIMPSON HURRICANE SCALE

Hurricanes that reach a 3 to 5 on the Saffir-Simpson Hurricane Scale (formulated in 1971 by American engineer Herbert Saffir and meteorologist Robert Simpson) are considered major.

Category 1—winds from 74 to 95 mph (119 to 153 km/h); minor to medium damages
Category 2—winds from 96 to 110 mph (154 to 177 km/h); considerable damages along the coast; trees uprooted
Category 3—winds from 111 to 130 mph (178 to 209 km/h); severe damage to small structures along the coast; roofs blown off
Category 4—winds from 131 to 155 mph (210 to 249 km/h); severe damage along the coast and inland
Category 5—winds greater than 156 mph (250 km/h); rare phenomenon; buildings may collapse

## APPENDIX 4: THE FIVE GREAT EXTINCTIONS OF LIFE ON EARTH

Ordovician–Silurian extinction: 444–447 million years ago
Late Devonian extinction: 364 million years ago
Permian–Triassic extinction: 251 million years ago
Triassic–Jurassic extinction: 200 million years ago
Cretaceous–Tertiary extinction: 65.5 million years ago

## APPENDIX 5: MAJOR (OFFICIAL) NUCLEAR ACCIDENTS SINCE 1952

This list is partially taken from *Les Jeux de l'atome et du hasard* [Games of the Atom and Chance] by Jean-Pierre Pharabod and Jean-Paul Schapira, published by Calmann-Lévy in 1988. It may seem long (although far from exhaustive), but this list is necessary to allow the reader to fully comprehend the risks we run by relying on nuclear energy. On a broader scale, the nuclear example reflects the scope of the hazards to which public well-being is exposed by political and industrial lobbyists (as is also the case with GMOs, oil, household cleaning products, etc.).

### December 12, 1952—Chalk River (Canada)

Following an operator error on the control rods, thirty-one employees are exposed to radiation levels ranging from 40 to 170 mSv (millisieverts).

### November 29, 1955—National Reactor Testing Station (Idaho)

An operator error on the control rods leads to a 40 to 50 percent core meltdown.

### October 1956—Marcoule (France)

As the graphite-gas military reactor reaches its maximum power for the first time, a channel's fuel oxidizes and melts due to an accidental reduction of the emission rate.

### October 8–12, 1957—Windscale (England)

A military plutonium production reactor catches fire. Significant contamination of part of Cumbria; weaker contamination of a large part of England. Most likely resulted in dozens of terminal cancers.

### May 24, 1958—Chalk River (Canada)

A fuel rod in the NRU heavy-water reactor is damaged during power escalation and catches fire while it is being emptied.

Significant contamination limited to the reactor building. A few employees are exposed to 200 mSv of radiation.

### October 18, 1958—Boris Kidric Institute (Yugoslavia)

The level of heavy water is inaccurately set, causing six people to be severely irradiated (one fatality).

### November 18, 1958—National Reactor Testing Station (Idaho)

Melting of HTRE-3's fuel element rings.

### December 14, 1959—Marcoule (France)

A sudden, extreme temperature rise in a channel goes undetected due to a wiring error. Despite unfavorable weather conditions, the reactor's $CO_2$ is emptied out in order to allow for repairs, leading to the irradiation of local inhabitants, who were apparently not warned. According to the authorities, the irradiation level is very low; however, personnel are exposed to severe irradiation during repairs.

### January 3, 1961—National Reactor Testing Station (Idaho)

Nuclear power excursion of the small SL-1 water reactor. Three dead, several people exposed to between 30 and 270 mSv of radiation.

### March 20, 1965—Chinon (France)

Despite signs barring entry, an employee enters an activated zone to retrieve something he forgot. He is exposed to 500 mSv of radiation.

**December 30, 1965—Mol (Belgium)**

Limited power excursion. One person severely irradiated (foot amputated).

**October 5, 1966—Laguna Beach (Michigan)**

Due to a piece of metal blocking sodium circulation, two structures holding fuel elements in the breeder reactor melt. For a month, engineers do not risk intervening for fear of creating a "critical mass" in the core.

**November 7, 1967—Grenoble (France)**

Fusion of a fuel element in the Siloe research reactor. 55,000 curies are released into the water tank and 2,000 curies into the atmosphere.

**January 21, 1969—Lucens (Switzerland)**

Sudden decompression of the primary cooling circuit for a 30-MWth reactor built in a cave. Severe contamination. The reactor is abandoned and the cave is sealed off.

**October 17, 1969—Saint-Laurent-Des-Eaux (France)**

Fusion of 110 pounds of uranium while a reactor is being loaded. Contamination is allegedly limited to the site.

**June 5, 1970—Dresden 2 (Illinois)**

Following an erroneous low-water level signal in the reactor, the operator sends in more water. One minute later, he realizes the signal was inaccurate but cannot completely interrupt the arrival of water. He halts the rise in pressure by opening a valve that partially floods the reactor building. The interior of the building is contaminated; two months are needed for repairs.

**September 1973—Chevtchenko (Kazakhstan, former USSR)**

Eight hundred eighty pounds of water enter the secondary sodium circuit (nonradioactive). There is a sodium-water explosion, a bursting disc ruptures, and hydrogen produced is released into the atmosphere and spontaneously combusts. American satellites detect the fire. The potential number of victims is unknown.

**November 7, 1973—Vermont Yankee Power Station (Vermont)**

During a core check, an accidental criticality takes place: A fuel rod was mistakenly left out of the core. The chain reaction is immediately and automatically brought to a halt by dropping the safety rod.

**May 2, 1974—Savannah River (South Carolina)**

Leaks in a military heavy-water plutonium production reactor. Surrounding areas are contaminated by tritium.

**July 19, 1974—Grenoble (France)**

Twenty-five hundred curies of radioactive antimony-124 leak into the reactor water tank. Due to excessive volumes of radioactive effluents flowing into insufficiently impermeable sewers, the water table is contaminated. In certain places, nine times the maximum acceptable concentration is recorded. The Central Service of Protection Against Ionizing Radiation (SCPRI) does not alert the population of Grenoble.

**August 20, 1974—Beznau (Switzerland)**

A 350-MWe pressurized water reactor undergoes the beginning of
what would trigger the accidental chain of events at Three Mile
Island, but after three minutes the operator realizes that the
pressurizer's discharge valve has remained open, and in nine
minutes the incident is brought under control. However, the
discharge balloon is ruptured, and the interior of the confinement
zone is lightly contaminated.

**February 1975—Chevtchenko (Kazakhstan,
former USSR)**

A massive amount of water—weighing 1,760 pounds—enters the
secondary sodium circuit. Deterioration of a vapor generator; 660
pounds of sodium catch fire.

**March 22, 1975—Brown's Ferry (Alabama)**

While working on the construction of a new reactor, a worker
attempts to check the cable room's overpressure with a candle,
setting a fire that spreads throughout the cabling and puts the
cooling system of the Unit 1 core out of commission. Operators
succeed in manually shutting down the reactor and cooling it with
the cooling circuit of the reactor that isn't operating (RRA). Unit 2
is also shut down.

**January 5, 1976—Bohunice (Slovakia, former
Czechoslovakia)**

During a fueling operation, the heavy-water reactor is suddenly
and accidentally depressurized and is cooled by $CO_2$ at 60 atm.
The gas asphyxiates two workers. Unreported amounts of
radiation are released outside the plant.

**March 28, 1979—Three Mile Island (Pennsylvania)**

Partial meltdown of the core of TMI-2 at Three Mile Island.
Damages total $1 billion.

**March 13, 1980—Saint-Laurent-Des-Eaux (France)**

Fuel overheats and two elements are totally fused when the reactor
builds to power too quickly. There is significant contamination in
the intervention zone. According to the SCPRI, irradiation of
neighboring inhabitants remains below maximum acceptable levels.

**September 23, 1983—Constituyentes (Argentina)**

The configuration of a reactor core is modified without following
safety guidelines, provoking a power excursion and the operator's
death by irradiation.

**April 14, 1984—Saint-Vulbas (France)**

Following a gradual decrease in direct voltage, the fuel rods drop
and the turbine is set in motion. The lack of direct voltage
requires the backup diesel generators to supply the reactor with
electricity. The first generator, which is connected to the
malfunctioning line, cannot be started. Happily, the second
generator starts up and enables the operations required to cool the
reactor. No other backup security system is in place.

**July 1, 1984—Saint-Laurent-Des-Eaux (France)**

An operator mistakenly orders the opening of the valves of Saint-
Laurent B2, which is running. The irruption of the primary water
would have broken the circuit and caused a significant loss of
coolant. Luckily, the valves did not function due to the difference

in pressure. This error is due to a money-saving measure that called for a single building to be an auxiliary for two reactors.

### April 26, 1986—Pripyat (Ukraine, former USSR)
The Chernobyl disaster, the consequences of which have gone down in history.

### January 12, 1987—Saint-Laurent-Des-Eaux (France)
Ice on the Loire halts the cooling of graphite-gas reactor Saint-Laurent 1, which must urgently be stopped. The diesel generators cannot cool the shut-down reactor, as they are also out of commission, and cooling relies on electricity from the EDF network for an hour. A few hours later, this network goes down in western France, including in Saint-Laurent; luckily, by this point the diesel generators are functioning.

*This list only accounts for the principal accidents disclosed between 1952 and 1988. A list of more recent accidents would have to include the following:*

### September 30, 1999—Tokaimura (Japan)
Introduction of fuel into a settling tank following an operator's error leads to a criticality reaction. The accident exposes more than six hundred local inhabitants to significant levels of radiation and kills two employees; it is considered the most severe nuclear accident after Chernobyl.

### February 15, 2000—Indian Point (New York)
The reactor at the Indian Point Nuclear Power Plant releases a small amount of radioactive vapor.

### July 2000—Washington (United States)
Wildfires near the highly radioactive Hanford Site for nuclear waste. No aerial contamination is detected beyond the site.

### August 9, 2004—Fukui (Japan)
Accident in the Mihama KEPCO power plant (five dead, six injured). The cause of the accident is a nonradioactive vapor leak in a building sheltering Mihama-3's turbines.

# bibliography

## BOOKS

Jean-Louis ÉTIENNE. *Le Pôle intérieur* [The Interior Pole]. Paris: Éditions J'ai Lu, 2001.

Nicolas HULOT. *Le Syndrome du Titanic* [The Titanic Syndrome]. Paris: Calmann-Lévy, 2004.

Edgar MORIN and Jean BAUDRILLARD. *La Violence du monde* [The Violence of the World]. Paris: Éditions du Félin, 2003.

Jean-Marie PELT. *La Terre en héritage* [Inheriting the Earth]. Paris: Fayard, 2001.

Jean-Pierre PHARABOD and Jean-Paul SCHAPIRA. *Les Jeux de l'atome et du Hasard* [Games of the Atom and Chance]. Paris: Calmann-Lévy, 1988.

Hubert REEVES. *Mal de Terre* [Earth Sickness], Science ouverte series. Paris: Éditions du Seuil, 2003.

## OFFICIAL STATEMENTS AND REPORTS

ACIA Overview Report. *The Arctic Climate Impact Assessment.* October 2004.

INSU/CNRS. (National Institute for Sciences of the Universe). "New data on greenhouse gases, 210,000 years earlier than previous data, found in the Dome C EPICA ice core." November 25, 2005.

Pentagon. *An Abrupt Climate Change Scenario and Its Implications for United States National Security.* Peter Schwartz and Doug Randall. October 2003.

French Senate Legislature. Xavier Pintar. May 25, 2000. *Draft for a law authorizing the Kyoto Protocol of the United Nations convention on climate change.* Session of Legislature.

Reports of the Intergovernmental Panel on Climate Change (IPCC).

U.N. special report on GMOs and the right to food.

## PERIODICALS

Boulanger, Olivier. "The 2003 Heat Wave, An Exceptional Meteorological Event?" *CSI Sciences actualités* [CSI Science News], October 2003.

Delmas, Juliette. "More and More Hurricanes." *Nature,* January 2004.

Gauthier, Philippe. "The Melting of the Glaciers Threatens India with Floods, then Drought." *Quebec Science,* November 1999.

"The Arctic Continues to Melt." *WWF,* November 2004.

## WEB SITES

National Center for Space Studies, "El Niño, a Challenge for Oceanographers." January 2004. www.cnes.fr.

Official site of The Environmental Science Published for Everybody Round the Earth. www.espere.net.

Official site of Canadian newspaper *Le Devoir.* www.ledevoir.com.

MétéoSuisse Federal Meteorology and Climatology Office. "Tropical Cyclones over the Atlantic in 2005—a Record-breaking Season." Lionel Peyraud, December 2005. www.meteosuisse.ch.

Canadian Weather Service. www.msc.ec.gc.ca.

## DOCUMENTARY FILM

Horner, Chris and Gilliane Le Gallic. *Nuages au Paradis* [Clouds in Paradise], 2004.

# acknowledgments

I would like to acknowledge those who helped me produce this book.

My deepest thanks to Jean-Marie Pelt, who has supported me from the beginning—for his encouragement over several years, for his generosity and advice, and for honoring me with the preface to this book.

Thank you to Florence Bastien, Anne Carton, Francis Carton, Gérard Lahache, Loic Lequere, Florian Martin, Jean-Claude Philbert, Franck Seigneur, Tony Travouillon, Peggy Vincent, and to the National Aeronautics and Space Administration (NASA) for photographs and advice.

This final picture was taken from the moon during one of the Apollo missions. It serves to remind us that our planet is absolutely unique in the universe. It is in our best interest to preserve this environment, for in case of trouble, we don't yet have anywhere else to run.

"Today's problems will only be resolved once we have resolved to concentrate on tomorrow."

# sources

René Barjavel, *Ravage.* © 1943 Denoël, Paris.

Jean Baudrillard and Edgar Morin, *La Violence du monde* [The Violence of the World]. © 2003 Éditions du Félin, Paris.

Émile Cioran, *De l'inconvénient d'être né* [Of the Inconvenience of Being Born]. © 1973 Éditions Gallimard, Paris.

Jean Dorst, *La Nature dénaturée* [Nature Denatured]. © 1965 Delachaux et Niestlé, Paris.

Albert Einstein, *Comment je vois le monde* [The World as I See It]. © 1979 Flammarion, Paris.

Jean-Louis Étienne, *Le Pôle intérieur* [The Interior Pole]. © 1999 Éditions Hoëbecke, Paris.

Nicolas Hulot, *Le Syndrome du Titanic* [The Titanic Syndrome]. © 2004 Éditions Calmann-Lévy, Paris.

Hubert Reeves, *Mal de Terre* [Earth Sickness]. © 2003 Seuil, Paris.

Stephen Jay Gould, *The Panda's Thumb: More Reflections in Natural History.* © 1980 Stephen Jay Gould / W.W. Norton, New York.

Robert Kandel, *Le Devenir des climats* [The Future of Climate]. © 1995 Hachette, Paris.

Théodore Monod, *Terre et Ciel* [Heaven and Earth]. © 1999 Actes Sud, Arles.

Jean Rostand, *Pensées d'un biologiste* [Thoughts of a Biologist]. © 1973 J'ai lu, Paris.

Jean Rostand, foreword in Édouard Bonnefous, *L'Homme ou la nature?* [Man or Nature?]. © 1970 Hachette, Paris.

Philippe Saint-Marc, *Socialisation de la nature* [Socialization of Nature]. © 1971 Stock, Paris.

Satprem, *La Révolte de la terre* [The Revolt of the Earth]. © 1990 Éditions Robert Laffont, Paris.